Quantum Many-Body Physics

Quantum Many-Body Physics: A Path Integral Approach offers a fresh perspective on quantum phenomena in condensed matter physics. This comprehensive first edition provides a powerful and versatile framework for understanding complex problems in systems such as ultracold atoms, superfluids, and superconductors.

The text begins with a solid foundation in canonical quantum field theory before progressing to advanced topics, ensuring readers develop both deep conceptual understanding and practical skills.

Features include:

- Concise yet thorough coverage of path integral methods.
- Practical applications to real-world quantum systems.
- Problem-solving techniques that bridge theory and experiment.

Designed for graduate students in physics and related disciplines, this book serves as an essential resource for those aiming to advance their studies in quantum mechanics. Researchers in condensed matter physics, quantum optics, and quantum information will also find valuable insights to support their work on the next breakthroughs in the field.

Luca Salasnich is a full Professor in Theoretical Physics of Matter at the Department of Physics and Astronomy "Galileo Galilei" of the University of Padova. His fields of research are condensed matter theory and statistical physics. In particular, he studies nonlinear phenomena (chaos, solitons, and vortices) and macroscopic quantum effects (Bose-Einstein condensation, superfluidity, and superconductivity) in ultracold atomic gases and other many-body systems by using analytical and numerical methods.

Textbook Series in Physical Sciences

This textbook series offers pedagogical resources for the physical sciences. It publishes high-quality, high-impact texts to improve understanding of fundamental and cutting edge topics, as well as to facilitate instruction. The authors are encouraged to incorporate numerous problems and worked examples, as well as making available solutions manuals for undergraduate and graduate level course adoptions. The format makes these texts useful as professional self-study and refresher guides as well. Subject areas covered in this series include condensed matter physics, quantum sciences, atomic, molecular, and plasma physics, energy science, nanoscience, spectroscopy, mathematical physics, geophysics, environmental physics, and so on, in terms of both theory and experiment.

Understanding Nanomaterials
Malkiat S. Johal

Concise Optics: Concepts, Examples, and Problems
Ajawad I. Haija, M. Z. Numan, W. Larry Freeman

A Mathematica Primer for Physicists
Jim Napolitano

Understanding Nanomaterials, Second Edition
Malkiat S. Johal, Lewis E. Johnson

Physics for Technology, Second Edition: With Applications in Industrial Control Electronics
Daniel H. Nichols

Time-Resolved Spectroscopy: An Experimental Perspective
Thomas Weinacht; Brett J. Pearson

No-Frills Physics: A Concise Study Guide for Algebra-Based Physics
Matthew D. McCluskey

Quantum Principles and Particles, Second Edition
Walter Wilcox

Electronic Conduction: Classical and Quantum Theory to Nanoelectronic Devices
John P. Xanthakis

Physics: A Focused Introduction
Colin Campbell

Quantum Many-Body Physics: A Path Integral Approach
Luca Salasnich

For more information about this series, please visit: [www.crcpress.com/Textbook-Series-in-Physical-Sciences/book-series/TPHYSCI]

Quantum Many-Body Physics

A Path Integral Approach

Luca Salasnich

CRC Press
Taylor & Francis Group
Boca Raton London New York

CRC Press is an imprint of the
Taylor & Francis Group, an **informa** business

First edition published 2026
by CRC Press
2385 NW Executive Center Drive, Suite 320, Boca Raton FL 33431

and by CRC Press
4 Park Square, Milton Park, Abingdon, Oxon, OX14 4RN

CRC Press is an imprint of Taylor & Francis Group, LLC

ISBN: 978-1-032-96134-7 (hbk)
ISBN: 978-1-041-15190-6 (pbk)
ISBN: 978-1-003-58833-7 (ebk)

DOI: 10.1201/9781003588337

Typeset in CMR12 font
by KnowledgeWorks Global Ltd.

Publisher's note: This book has been prepared from camera-ready copy provided by the authors.

Contents

Preface

This book contains lecture notes prepared for the course "Theory of Strongly Correlated Systems" offered by the author at the University of Padova for the MSc in Physics. The aim of the course is to discuss relevant quantum phenomena in condensed matter and atomic physics by means of the path integral approach.

The path integral formulation is useful, especially for analyzing many-body quantum systems. Here are the main benefits: i) Path integrals unify quantum mechanics and statistical mechanics. By summing over all possible paths in both position and time, it extends to finite temperature systems via the Wick rotation and the Matsubara formalism, making it suitable for both ground-state and finite-temperature problems. ii) Path integrals are good for complex interactions, like in many-body systems. This approach can be used for strongly interacting systems, where traditional methods may fail. iii) The path integral approach makes quantum mechanics easier to understand. Instead of dealing with wave functions, it introduces the idea of summing over all possible paths. This helps in understanding concepts like quantum tunneling, quantum coherence, and the effect of fluctuations. iv) Many condensed matter systems, like superconductors and quantum fluids, are described by quantum field theory (QFT). Path integrals are a useful tool for deriving and studying QFT models of condensed matter systems. These models describe many-body phenomena like superconductivity, magnetism, and quantum criticality. v) Path integrals can be used to study non-equilibrium systems. These methods are useful for describing systems that are not at equilibrium, such as those changing rapidly or subject to external fields. The path integral formalism makes it easy to include time-dependent effects in the analysis. vi) Path integrals are useful in numerical simulations, especially Quantum Monte

Carlo methods. It is often difficult to solve condensed matter problems analytically. However, path integrals allow for efficient Monte Carlo sampling techniques to approximate the behavior of systems, especially for high-dimensional problems. vii) Path integrals can include topological effects, which are very important in many condensed matter systems. In particular, path integrals can be used to study the behavior of topological phases and quantum phase transitions. viii) In condensed matter physics, effective low-energy theories are often derived from more fundamental interactions. The path integral approach is a powerful tool for deriving such effective theories, especially when integrating high-energy degrees of freedom. ix) The path integral formulation makes it easy to include external fields in the system. The coupling to fields can be treated straightforwardly by modifying the action, making it a versatile tool for problems involving charged particles or systems in the presence of external forces. x) Path integrals can capture non-perturbative effects, such as instantons, solitons, quantized vortices, and other topological excitations, which are often difficult to address using traditional perturbation theory. These effects are especially relevant in strongly correlated systems and are critical for understanding phenomena like phase transitions, quantum criticality, and the behavior of strongly interacting quantum systems. In conclusion, the path integral approach represents a robust and adaptable methodology within the field of condensed matter physics. It offers a systematic approach to investigating both equilibrium and non-equilibrium systems, incorporates quantum fluctuations, and enables the numerical simulation of intricate many-body problems.

This book is organized into five chapters. The first chapter reviews the canonical quantization in nonrelativistic quantum mechanics and optics. The chapter starts with a discussion of the least action principle for classical systems and then it faces the canonical quantization of particles, bosonic fields, and fermionic fields. In this chapter are also discussed coherent states and Grassmann variables which are crucial mathematical tools for developing the path integral approach to quantum fields. The second chapter shows that it is possible to describe the quantum mechanics of particles by using functional integrals. In this chapter several applications of the Feynman path integral are considered. Among them, the free particle, the particle in a harmonic potential, the quantum tunneling, and the deep connection to finite-temperature problems by using the Wick rotation. The third chapter considers the quantum statistical mechanics of bosons within the formalism of functional integration for the bosonic field, introducing the equation of the state of weakly interacting bosons and the dimensional regularization of Gaussian fluctuations. It also analyzes the time-dependent Gross-Pitaevskii equation: its connection to the equations of superfluid hydrodynamics of ultracold atoms and its topological and solitonic solutions (quantized vortices, dark solitons, and bright solitons). The fourth chapter introduces the formalism of functional integration for the fermionic field. It is then derived the equation of state of weakly-interacting repulsive fermions and the paring of attractive fermions within the mean-field Bardeen-Cooper-Schrieffer (BCS) theory of low-temperature superconductivity. A section is devoted to the phenomenological Ginzburg-Landau theory of superconductivity which is quite successful near the critical temperature for any kind of superconductor and, also for this reason, it is a paradigmatic and relevant application of functional integration. Finally, within the path-integral framework, we analyze the physics of Josephson junctions, which are the key ingredients of superconducting quantum computers, and also the BCS-BEC crossover from the BCS regime of atomic Cooper pairs to the Bose-Einstein condensation (BEC) of molecules. The last chapter of the book explains the formalism of functional integration for the electromagnetic field in the Coulomb gauge. By using the path integral it is then discussed the Planck law of black body radiation, the Casimir effect, the thermodynamics of photons in nonlinear Kerr media, and the BEC of photons in microcavities. At the end of each chapter there are solved problems, which are useful to better understand the subtleties of path integral.

For all the topics discussed in this book, the path integral formulation could be developed, without any reference to the second quantization operators, starting directly from the "classical" action of the quantum problem. The "classical" action will be characterized by real fields, complex fields, or Grassmann fields, depending on the problem. In this way there are no technical issues related to the zero point energy and there is no need to introduce convergence factors. This approach, usually adopted in high-energy physics, is discussed in chapters three and four, and compared with the condensed-matter perspective, which involves quantum coherent states to build the functional integral from ladder operators.

The author thanks Alberto Ambrosetti, Sandro Azaele, Andrea Bardin, Fiorenzo Bastianelli, Giacomo Bighin, Alberto Cappellaro, Luca Dell'Anna, Ganmaria Falasco, Piero Faccioli, Koichiro Furutani, Giacomo Gradenigo, Serghei Klimin, Francesco Lorenzi, Pieralberto Marchetti, Roberto Onofrio, Fabio Sattin, Jacques Tempere, Flavio Toigo, Andrea Tononi, and Cesare Vianello for many suggestions and enlightening discussions about path integrals in quantum mechanics, quantum field theory, and stochastic processes.

Padova, May, 2025 Luca Salasnich

1

Canonical Quantum Field Theory

In this chapter we review the main ideas of the canonical quantization in nonrelativistic quantum mechanics and optics, which we suppose the reader learned in previous courses. We first consider the least action principle and the action functional for classical many-body systems and also for the Schrödinger wavefunction, interpreted as the complex field of a classical field theory. Then, after having analyzed the Legendre transformation from the Lagrangian to the Hamiltonian formulation, we illustrate the canonical quantization of particles, and finally we extend the approach to the canonical quantization of bosonic and fermionic fields, considering also the role of temperature.

1.1 Introduction

Modern physics is Galilean science par excellence. It is in fact based on the experimental method and the mathematical description of natural phenomena. The success of physics is due to the reproducibility of experiments, the ability to make accurate predictions, and the many technological applications: just think of electronics or telecommunications with radio waves. Despite the great success of physics, the concepts of determinism and reductionism often occur in discussions between physicists. In the case of determinism, it is the question of understanding whether physical reality operates in a deterministic way proceeding from causes to effects or whether there are intrinsically random aspects. At the

DOI: 10.1201/9781003588337-1

beginning of the 20th century the determinism of classical mechanics underwent a heavy downsizing following the surprising experimental results of atomic physics, which led to the formulation of the quantum theory of matter. But there is more. In the second half of the 20th century, thanks to the help of electronic calculators, it was discovered that perfectly deterministic physical laws, such as those of Newtonian mechanics, can produce a completely chaotic motion. The issue about the reductionism is about understanding if the physical reality can be known breaking it down into its parts or if the "everything" is instead not reducible to the simple sum of its parts. In recent years, reductionism has been questioned, mainly in biology and chemistry, but also in physics. The theory of complexity is a modern trend of thought that believes that "everything" contains more information of the "parts" that constitute it. However, this is not a true abandonment of reductionism but, more pragmatically, the observation that some phenomena, called collective phenomena, can only emerge as a result of the interaction of "parts" strongly correlated. This fact is well known in statistical physics and condensed matter, as we will show by analyzing phase transitions and phenomena such as superfluidity and superconductivity. Contrary to what they also claim authoritative intellectuals, the theory of complexity does not imply a holistic vision or pantheism of reality. The theory of complexity teaches us that simple systems can have complex behavior and that complex systems can have simple behavior.

This book is about the complex behavior of quantum many-body systems. It is, however, important to start by reminding the basic properties of classical and quantum mechanics. This chapter is dedicated to illustrate these physical-mathematical tools while also providing the historical context in which they were formulated.

1.2 Least Action Principle

In the 1750s Leonhard Euler and Giuseppe-Luigi Lagrange discovered that the classical mechanics of Newton can be reformulated within a variational principle: the principle of least action. Let us consider, for the sake of simplicity, a particle of mass m with coordinate $x(t)$ at time t described by the Lagrangian

$$L = \frac{1}{2} m \dot{x}^2 - U(x) \,, \tag{1.1}$$

where

$$\dot{x}(t) = \frac{d}{dt} x(t) \tag{1.2}$$

is the velocity of the particle and $U(x)$ is a generic external potential acting on the particle. The corresponding action functional is defined as

$$S[x] = \int_{t_I}^{t_F} dt \, L \,, \tag{1.3}$$

where t_I is the initial time and t_F is the final time. Notice that, in general, the Lagrangian is not univocal, but is defined up to any function that can be written as a derivative with respect to time.

Considering an infinitesimal variation $\delta x(t)$ of $x(t)$ at fixed boundaries, i.e., imposing that $\delta x(t_I) = \delta x(t_F) = 0$, one obtains an infinitesimal variation δS of the action which can be formally written as

$$\delta S = \int_{t_I}^{t_F} dt \, \frac{\delta S[x]}{\delta x(t)} \, \delta x(t) \,, \tag{1.4}$$

where the functional derivative reads

$$\frac{\delta S[x]}{\delta x(t)} = \frac{\partial L}{\partial x} - \frac{d}{dt} \frac{\partial L}{\partial \dot{x}} \,. \tag{1.5}$$

To prove this relevant result, we set $L = L(x, \dot{x})$ and perform a Taylor expansion, finding

$$\begin{aligned} S[x + \delta x] &= \int_{t_I}^{t_F} dt \, L(x + \delta x, \dot{x} + \dot{\delta x}) \\ &= \int_{t_I}^{t_F} dt \left\{ L(x, \dot{x}) + \frac{\partial L(x, \dot{x})}{\partial x} \delta x + \frac{\partial L(x, \dot{x})}{\partial \dot{x}} \dot{\delta x} \right\} \end{aligned} \tag{1.6}$$

neglecting terms more than linear with respect to δx. Thus,

$$S[x + \delta x] = S[x] + \delta S[x] \,, \tag{1.7}$$

where

$$\begin{aligned}\delta S[x] &= \int_{t_I}^{t_F} dt \left\{ \frac{\partial L(x,\dot{x})}{\partial x}\delta x + \frac{\partial L(x,\dot{x})}{\partial \dot{x}}\dot{\delta x} \right\} \\ &= \int_{t_I}^{t_F} dt \left\{ \left(\frac{\partial L(x,\dot{x})}{\partial x} - \frac{d}{dt}\frac{\partial L(x,\dot{x})}{\partial \dot{x}} \right) \delta x \right\} \\ &+ \left[\frac{\partial L(x,\dot{x})}{\partial \dot{x}} \delta x \right]_{t_I}^{t_F} \end{aligned} \tag{1.8}$$

having used integration by parts. As previously anticipated, imposing the variation $\delta x(t)$ vanishes at the boundaries, i.e., $\delta x(t_I) = \delta x(t_F) = 0$, the last term of Eq. (1.8) is zero, and we obtain exactly Eq. (1.5).

The least action principle, that is actually a theorem, says that the extremization of the action $S[x]$, namely

$$\delta S = 0 \ , \tag{1.9}$$

gives the so-called Euler-Lagrange equation

$$\frac{\partial L}{\partial x} - \frac{d}{dt}\frac{\partial L}{\partial \dot{x}} = 0 \ . \tag{1.10}$$

This result is indeed a direct consequence of Eqs. (1.5) and (1.9). For dynamically stable trajectories $x(t)$, the extremization $\delta S[x(t)] = 0$ corresponds to a minimization. This fact can be proved by analyzing the second variation $\delta^2 S[x(t)]$ of the action, which must be positive to have dynamical stability. Historically, a very simple version of the action principle was formulated for the first time by Pierre Louis Maupertuis in 1744. The word "action" was chosen with the idea that "God wants to minimize its action on the Universe". Taking into account Eq. (1.1), we explicitly find

$$\frac{\partial L}{\partial x} = -\frac{\partial U}{\partial x} \ , \tag{1.11}$$

$$\frac{\partial L}{\partial \dot{x}} = m\dot{x} \ . \tag{1.12}$$

Then the Euler-Lagrange equation (1.10) becomes

$$m\ddot{x} = -\frac{\partial U}{\partial x} \ , \tag{1.13}$$

that is nothing else than the Newton law for the dynamics of the particle with acceleration

$$\ddot{x}(t) = \frac{d^2}{dt^2}x(t) = a(t) \tag{1.14}$$

induced by the conservative force

$$F(x) = -\frac{\partial U(x)}{\partial x} \ , \tag{1.15}$$

i.e., the familiar formula $m\,a = F$. Notice that the Cauchy problem, namely Eq. (1.13) with initial conditions $x(t_I) = x_I$ and $\dot{x}(t_I) = v_I$, has quite generically only one real solution. The same holds for the boundary problem, namely Eq. (1.13) with the conditions $x(t_I) = x_I$ and $x(t_F) = x_F$ in the case of a free particle and in the case of linear or quadratic potential $U(x)$. However, for any potential $U(x)$ rising faster than quadratically at large positive and negative x, Eq. (1.13) with boundary conditions $x(t_I) = x_I$ and $x(t_F) = x_F$ admits several real solutions.

1.2.1 Many-Body Action

Quite generally, the non-dissipative dynamical equations of classical physics can be obtained with the least action principle having identified the appropriate action functional and Lagrangian function. For instance, for a system of N identical particles of mass m characterized by the position vectors $\mathbf{r}_i(t) = (x_i(t), y_i(t), z_i(t))$ and the vector velocities

$$\dot{\mathbf{r}}_i(t) = \frac{d}{dt}\mathbf{r}_i(t) \ , \tag{1.16}$$

with i=1,2,...,N, the many-body Lagrangian is given by

$$L = \sum_{i=1}^{N} \left(\frac{1}{2} m \dot{\mathbf{r}}_i^2 - U(\mathbf{r}_i) \right) - \frac{1}{2} \sum_{\substack{i,j=1 \\ i \neq j}}^{N} V(\mathbf{r}_i, \mathbf{r}_j) \ , \tag{1.17}$$

where $U(\mathbf{r}_i)$ is the external potential acting on the i-th particle and $V(\mathbf{r}_i, \mathbf{r}_j)$ is the internal interaction potential between the i-th particle and the j-th particle, such that $V(\mathbf{r},\mathbf{r}) = V(\mathbf{r}',\mathbf{r})$ to respect the third principle of dynamics. In Eq. (1.17) both i and j run from 1 to N and the factor 1/2 in front of the double summation avoids double counting. Moreover, $i \neq j$ means that any particle is not self-interacting. The corresponding action reads

$$S[\{\mathbf{r}_i\}] = \int_{t_I}^{t_F} dt\, L(\{\mathbf{r}_i\}, \{\dot{\mathbf{r}}_i\}) \tag{1.18}$$

being a functional of all the coordinates, collectively denoted as $\{\mathbf{r}_i(t)\}$. Eq. (1.18) is the many-body action of N identical interacting particle. It is straightforward to prove that the extermination of this action gives rise to N Euler-Lagrange equations

$$\frac{\partial L}{\partial \mathbf{r}_i} - \frac{d}{dt}\frac{\partial L}{\partial \dot{\mathbf{r}}_i} = 0 \tag{1.19}$$

which can be explicitly written as

$$m\ddot{\mathbf{r}}_i = -\frac{\partial}{\partial \mathbf{r}_i}\left(U(\mathbf{r}_i) + \sum_{\substack{j=1\\ j\neq i}}^{N} V(\mathbf{r}_i,\mathbf{r}_j)\right) . \tag{1.20}$$

1.2.2 Schrödinger Action

Remarkably, also the time-dependent Schrödinger equation of the complex wavefunction $\psi(\mathbf{r},t)$ can be derived from a variational principle. In this scheme, $\psi(\mathbf{r},t)$ must be interpreted as the complex field of classical field theory. For the Schrödinger field $\psi(\mathbf{r},t)$ of a particle of mass m in the presence of an external potential $U(\mathbf{r})$, we introduce the following Lagrangian

$$L = \int d^3\mathbf{r}\, \mathscr{L} , \tag{1.21}$$

where

$$\mathscr{L} = \frac{i\hbar}{2}\left(\psi^*\partial_t\psi - \psi\partial_t\psi^*\right) - \frac{\hbar^2}{2m}|\boldsymbol{\nabla}\psi|^2 - U\,|\psi|^2 \tag{1.22}$$

is the so-called Lagrangian density of the Schrödinger scalar field with $i = \sqrt{-1}$ the imaginary unit, $\hbar$ the reduced Planck constant,

$\partial_t = \frac{\partial}{\partial t}$ the partial derivative with respect to time t, and $\boldsymbol{\nabla} = \partial_{\mathbf{r}} = \frac{\partial}{\partial \mathbf{r}} = (\partial_x, \partial_y, \partial_z)$ the Laplace operator of partial spatial derivatives. The Schrödinger action functional reads

$$S[\psi^*, \psi] = \int_{t_I}^{t_F} dt \int d^3\mathbf{r}\, \mathscr{L}(\psi^*, \psi) \;, \tag{1.23}$$

where we write $\mathscr{L} = \mathscr{L}(\psi, \psi^*)$ to emphasize that the Lagrangian density can be interpreted as a function of two fields: ψ and its complex conjugate ψ^*. Treating the variations $\delta\psi$ and $\delta\psi^*$ of the two fields as independent, we get

$$\delta S = \int_{t_I}^{t_F} dt \int d^3\mathbf{r} \left(\frac{\delta S}{\delta \psi}\, \delta\psi + \frac{\delta S}{\delta \psi^*}\, \delta\psi^* \right) \;, \tag{1.24}$$

where the functional derivatives are

$$\frac{\delta S}{\delta \psi} = \frac{\partial \mathscr{L}}{\partial \psi} - \partial_t \frac{\partial \mathscr{L}}{\partial(\partial_t \psi)} - \boldsymbol{\nabla} \cdot \frac{\partial \mathscr{L}}{\partial(\boldsymbol{\nabla}\psi)} \;, \tag{1.25}$$

$$\frac{\delta S}{\delta \psi^*} = \frac{\partial \mathscr{L}}{\partial \psi^*} - \partial_t \frac{\partial \mathscr{L}}{\partial(\partial_t \psi^*)} - \boldsymbol{\nabla} \cdot \frac{\partial \mathscr{L}}{\partial(\boldsymbol{\nabla}\psi^*)} \;, \tag{1.26}$$

provided that the variations $\delta\psi(\mathbf{r}, t)$ and $\delta\psi^*(\mathbf{r}, t)$ vanish at the space-time boundary conditions. Imposing that $\delta S = 0$, we obtain the Euler-Lagrange equations

$$\frac{\partial \mathscr{L}}{\partial \psi} - \partial_t \frac{\partial \mathscr{L}}{\partial(\partial_t \psi)} - \boldsymbol{\nabla} \cdot \frac{\partial \mathscr{L}}{\partial(\boldsymbol{\nabla}\psi)} = 0 \;, \tag{1.27}$$

$$\frac{\partial \mathscr{L}}{\partial \psi^*} - \partial_t \frac{\partial \mathscr{L}}{\partial(\partial_t \psi^*)} - \boldsymbol{\nabla} \cdot \frac{\partial \mathscr{L}}{\partial(\boldsymbol{\nabla}\psi^*)} = 0 \;. \tag{1.28}$$

After some calculations one finds that Eq. (1.28) gives the time-dependent Schrödinger equation

$$i\hbar \frac{\partial}{\partial t} \psi(\mathbf{r}, t) = \left(-\frac{\hbar^2}{2m} \nabla^2 + U(\mathbf{r}) \right) \psi(\mathbf{r}, t) \;, \tag{1.29}$$

while Eq. (1.27) gives the complex conjugate of the Schrödinger equation, namely

$$-i\hbar \frac{\partial}{\partial t} \psi^*(\mathbf{r}, t) = \left(-\frac{\hbar^2}{2m} \nabla^2 + U(\mathbf{r}) \right) \psi^*(\mathbf{r}, t) \;. \tag{1.30}$$

It is important to stress that often in the scientific literature it is used a slightly different Schrödinger Lagrangian density with respect to the one of Eq. (1.22). This alternative Lagrangian density is given by

$$\mathscr{L} = i\hbar\psi^*\partial_t\psi + \frac{\hbar^2}{2m}\psi^*\nabla^2\psi - U\,|\psi|^2\,. \tag{1.31}$$

Contrary to (1.22), this Lagrangian density does not seem like a real function. However, taking into account that

$$-\psi^*\partial_t\psi = \psi\partial_t\psi^* + \partial_t\left(|\psi|^2\right) \tag{1.32}$$

and also

$$-|\boldsymbol{\nabla}\psi|^2 = \psi^*\nabla^2\psi - \boldsymbol{\nabla}\cdot\left(\psi^*\boldsymbol{\nabla}\psi\right), \tag{1.33}$$

one easily proves that (1.22) is equivalent to (1.31) if the complex Schrödinger field $\psi(\mathbf{r},t)$ goes to zero at the space-time boundaries.

The Schrödinger Lagrangian and the Schrödinger action play a crucial role in the quantum field theory of nonrelativistic strongly-correlated systems. In this context it is necessary to include the self-interaction of the scalar field $\psi(\mathbf{r},t)$ that is mediated by the interaction potential $V(\mathbf{r},\mathbf{r}')$ as follows

$$\begin{aligned}\mathscr{L} &= \frac{i\hbar}{2}\left(\psi^*\partial_t\psi - \psi\partial_t\psi^*\right) - \frac{\hbar^2}{2m}|\boldsymbol{\nabla}\psi|^2 - U\,|\psi|^2 \\ &- \frac{1}{2}\int d^3\mathbf{r}'|\psi(\mathbf{r}',t)|^2V(\mathbf{r},\mathbf{r}')|\psi(\mathbf{r},t)|^2\,. \end{aligned} \tag{1.34}$$

The last term, which is nonlocal, plays a fundamental role in nonrelativistic field theory because, as we will see, it describes the interaction between identical particles of mass m, which are modeled collectively by the complex scalar field $\psi(\mathbf{r},t)$. The Euler-Lagrange equation obtained from this Lagrangian density is

$$i\hbar\frac{\partial}{\partial t}\psi(\mathbf{r},t) = \left(-\frac{\hbar^2}{2m}\nabla^2 + U(\mathbf{r}) + \int d^3\mathbf{r}'|\psi(\mathbf{r}',t)|^2V(\mathbf{r},\mathbf{r}')\right)\psi(\mathbf{r},t)\,, \tag{1.35}$$

that is a time-dependent Schrödinger equation with a nonlocal nonlinearity.

The presence of a nonlocal term makes a bit more difficult the determination of the functional derivatives. As a simple example, let us consider the functional

$$F[\phi^*, \phi] = \frac{1}{2} \int dx\, dx' \, |\phi(x)|^2 V(x, x') |\phi(x')|^2 \tag{1.36}$$

of the complex function $\phi(x)$ with $x \in \mathbb{R}$ and $V(x, x') = V(x', x)$. A small fluctuation $\delta\phi(x)$ of $\phi(x)$ gives

$$\begin{aligned} F[\phi^* + \delta\phi^*, \phi + \delta\phi] &= \frac{1}{2} \int dx\, dx' \, V(x, x') \\ & \times \Big\{ \left(\phi^*(x) + \delta\phi^*(x)\right) \left(\phi(x) + \delta\phi(x)\right) \\ & \left(\phi^*(x') + \delta\phi^*(x')\right) \left(\phi(x') + \delta\phi(x')\right) \Big\} \\ &= F[\phi] + \frac{1}{2} \int dx\, dx' \, V(x, x') \\ & \times \Big\{ |\phi(x')|^2 \phi^*(x) \delta\phi(x) \\ &+ |\phi(x')|^2 \phi(x) \delta\phi^*(x) + |\phi(x)|^2 \phi^*(x') \delta\phi(x') \\ &+ |\phi(x)|^2 \phi(x') \delta\phi^*(x') \Big\} + O(\delta\phi^2) \,, \end{aligned} \tag{1.37}$$

where $O(\delta\phi^2)$ means terms of order $\delta\phi^2$ or higher. By definition we have

$$F[\phi^* + \delta\phi, \phi + \delta\phi] = F[\phi^*, \phi] + \delta F[\phi^*, \phi] + \ldots \,, \tag{1.38}$$

where $\delta F[\phi^*, \phi]$ is the first variation of the functional $F[\phi^*, \phi]$. In our case we find

$$\delta F = \int dx\, dx' \, V(x, x') \Big\{ |\phi(x')|^2 \phi^*(x) \delta\phi(x) + |\phi(x')|^2 \phi(x) \delta\phi^*(x) \Big\} \tag{1.39}$$

after performing an exchange of the variables x and x' and taking into account that $V(x, x') = V(x', x)$. Finally, we write

$$\delta F = \int dx \Big\{ \frac{\delta F}{\delta\phi} \delta\phi(x) + \frac{\delta F}{\delta\phi^*} \delta\phi^*(x) \Big\} \,, \tag{1.40}$$

where

$$\frac{\delta F}{\delta \phi} = \int dx' \, V(x,x')|\phi(x')|^2 \phi^*(x) \quad (1.41)$$
$$\frac{\delta F}{\delta \phi^*} = \int dx' \, V(x,x')|\phi(x')|^2 \phi(x) \quad (1.42)$$

are the functional derivatives of $F[\phi]$.

1.2.3 Maxwell Action

We now analyze the free electromagnetic field in a vacuum, characterized by the electric field $\mathbf{E}(\mathbf{r},t)$ and the magnetic field $\mathbf{B}(\mathbf{r},t)$. Its Lagrangian can be written as in Eq. (1.21) in terms of a Lagrangian density $\mathscr{L}$. This Lagrangian density is formally given by

$$\mathscr{L} = \frac{\epsilon_0}{2}\mathbf{E}^2 - \frac{1}{2\mu_0}\mathbf{B}^2 \quad (1.43)$$

with ϵ_0 the dielectric constant in the vacuum and μ_0 the paramagnetic constant in the vacuum. Remember that $c = 1/\sqrt{\epsilon_0\mu_0} = 3 \times 10^8$ meters/seconds. The electromagnetic fields $\mathbf{E}(\mathbf{r},t)$ and $\mathbf{B}(\mathbf{r},t)$ can be written in terms of electromagnetic potentials as follows

$$\mathbf{E} = -\nabla\Phi - \frac{\partial \mathbf{A}}{\partial t} , \quad (1.44)$$
$$\mathbf{B} = \nabla \wedge \mathbf{A} , \quad (1.45)$$

where $\Phi(\mathbf{r},t)$ is the electromagnetic scalar potential and $\mathbf{A}(\mathbf{r},t)$ is the electromagnetic vector potential. Thus, the electromagnetic Lagrangian density can be expressed as

$$\mathscr{L} = \frac{\epsilon_0}{2}(\nabla\Phi + \partial_t\mathbf{A})^2 - \frac{1}{2\mu_0}(\nabla \wedge \mathbf{A})^2 . \quad (1.46)$$

In terms of the two electromagnetic potentials the action functional reads

$$S[\Phi,\mathbf{A}] = \int dt \int d^3\mathbf{r} \, \mathscr{L}(\Phi,\mathbf{A}) . \quad (1.47)$$

The minimization of this action functional gives rise to the Euler-Lagrange equations of the electromagnetic Lagrangian with respect to the scalar potential $\Phi(\mathbf{r}, t)$ and the vector potential $\mathbf{A}(\mathbf{r}, t)$:

$$\frac{\partial \mathscr{L}}{\partial \Phi} - \frac{\partial}{\partial t}\frac{\partial \mathscr{L}}{\partial(\partial_t \Phi)} - \boldsymbol{\nabla} \cdot \frac{\partial \mathscr{L}}{\partial(\boldsymbol{\nabla}\Phi)} = 0\,, \tag{1.48}$$

$$\frac{\partial \mathscr{L}}{\partial \mathbf{A}} - \frac{\partial}{\partial t}\frac{\partial \mathscr{L}}{\partial(\partial_t \mathbf{A})} - \boldsymbol{\nabla} \wedge \frac{\partial \mathscr{L}}{\partial(\boldsymbol{\nabla} \wedge \mathbf{A})} = \mathbf{0}\,. \tag{1.49}$$

They are nothing else than two of the four Maxwell equations in the absence of electric sources, namely

$$\boldsymbol{\nabla} \cdot \mathbf{E} = 0\,, \tag{1.50}$$

$$\boldsymbol{\nabla} \wedge \mathbf{B} = \epsilon_0 \mu_0 \frac{\partial}{\partial t}\mathbf{E}\,. \tag{1.51}$$

Note that the other two Maxwell equations are derived directly from Eqs. (1.44) and (1.45). They are

$$\boldsymbol{\nabla} \cdot \mathbf{B} = 0\,, \tag{1.52}$$

$$\boldsymbol{\nabla} \wedge \mathbf{E} = -\frac{\partial \mathbf{B}}{\partial t}\,. \tag{1.53}$$

Eqs. (1.44) and (1.45) do not determine the electromagnetic potentials uniquely. In fact, for an arbitrary scalar function $\Lambda(\mathbf{r}, t)$, the gauge transformation

$$\phi \to \phi' = \phi + \frac{\partial \Lambda}{\partial t}\,, \tag{1.54}$$

$$\mathbf{A} \to \mathbf{A}' = \mathbf{A} - \nabla\Lambda\,, \tag{1.55}$$

leaves the fields $\mathbf{E}$ and $\mathbf{B}$ unaltered. There is thus an infinite number of different electromagnetic potentials that correspond to a given configuration of measurable fields. We use this property to choose a gauge transformation such that

$$\nabla \cdot \mathbf{A} = 0\,. \tag{1.56}$$

This condition defines the Coulomb (or radiation) gauge, and the vector field $\mathbf{A}$ is called the transverse field. For a complex monochromatic plane wave

$$\mathbf{A}(\mathbf{r}, t) = \mathbf{A}_0\, e^{i(\mathbf{k}\cdot\mathbf{r} - \omega t)} \tag{1.57}$$

the Coulomb gauge (1.56) gives

$$\mathbf{k} \cdot \mathbf{A} = 0 \,, \tag{1.58}$$

i.e., $\mathbf{A}$ is perpendicular (transverse) to the wavevector $\mathbf{k}$. In the vacuum and without sources, under the Coulomb gauge (1.56) one gets

$$\phi(\mathbf{r}, t) = 0 \,. \tag{1.59}$$

As a consequence

$$\mathbf{E} = -\frac{\partial \mathbf{A}}{\partial t} \,, \tag{1.60}$$

$$\mathbf{B} = \boldsymbol{\nabla} \wedge \mathbf{A} \,. \tag{1.61}$$

Thus, in the Coulomb gauge one needs only the electromagnetic vector potential $\mathbf{A}(\mathbf{r}, t)$ to obtain the electromagnetic field. In this case the Lagrangian density becomes

$$\mathscr{L} = \frac{\epsilon_0}{2}(\partial_t \mathbf{A})^2 - \frac{1}{2\mu_0}(\boldsymbol{\nabla} \wedge \mathbf{A})^2 \,. \tag{1.62}$$

1.3 Lagrangian vs Hamiltonian

In this section we show how to move from the Lagrangian to the Hamiltonian formulation. We start again with the simple problem of a single particle described by the Lagrangian of Eq. (1.1). We introduce the canonical linear momentum

$$p = \frac{\partial L}{\partial \dot{x}} = m\dot{x} \tag{1.63}$$

and the Legendre transform

$$H = p\dot{x} - L(x, \dot{x}) \tag{1.64}$$

which defines the Hamiltonian H of the problem if one substitutes $\dot{x}$ with p, namely

$$H = p\frac{p}{m} - L(x, \dot{x} = \frac{p}{m}) \,. \tag{1.65}$$

Explicitly we have

$$H = \frac{p^2}{m} - \left(\frac{p^2}{2m} - U(x)\right) \tag{1.66}$$

and consequently

$$H = \frac{p^2}{2m} + U(x) \,. \tag{1.67}$$

This Hamiltonian is a function of the coordinate x and of the linear momentum p, i.e., $H = H(x,p)$. It represents the total energy of the system, that is the sum of the kinetic term $p^2/(2m)$ and the potential energy term $U(x)$. The inverse Legendre transformation

$$L = p\dot{x} - H(x,p) \tag{1.68}$$

can be used to write the action functional (1.3) in terms of the Hamiltonian. We immediately find

$$S[x,p] = \int_{t_I}^{t_F} dt\ (p\dot{x} - H(x,p)) = \int_{x(t_I)}^{x(t_F)} dx\, p - \int_{t_I}^{t_F} dt\, H(x,p) \,. \tag{1.69}$$

It follows that the first variation of $S[x,p]$ with respect to the two independent variables $x(t)$ and $p(t)$, with fixed ends for the coordinate, gives

$$\delta S = \int_{t_I}^{t_F} dt \left\{ \frac{\delta S}{\delta x}\delta x + \frac{\delta S}{\delta p}\delta p \right\} \,, \tag{1.70}$$

where the functional derivatives are

$$\frac{\delta S}{\delta x} = \dot{p} + \frac{\partial H}{\partial x} \,, \tag{1.71}$$

$$\frac{\delta S}{\delta p} = \dot{x} - \frac{\partial H}{\partial p} \,. \tag{1.72}$$

Imposing the least action principle $\delta S = 0$, we finally obtain

$$\dot{x} = \frac{\partial H}{\partial p} \,, \tag{1.73}$$

$$\dot{p} = -\frac{\partial H}{\partial x} \,, \tag{1.74}$$

which are the so-called Hamilton equations of motion. Using Eq. (1.67) they become

$$\dot{x} = \frac{p}{m} , \tag{1.75}$$

$$\dot{p} = -\frac{\partial U}{\partial x} . \tag{1.76}$$

These two first-order differential equations are equivalent to the second-order differential equation (1.13). Quite generally, for conservative systems, the Hamiltonian H is a constant of motion, i.e.

$$\frac{dH}{dt} = 0 . \tag{1.77}$$

This can be easily proved in our simple problem because

$$\frac{dH}{dt} = \frac{\partial H}{\partial t} + \frac{\partial H}{\partial p}\frac{dp}{dt} + \frac{\partial H}{\partial x}\frac{dx}{dt} = \dot{x}\dot{p} - \dot{p}\dot{x} = 0 \tag{1.78}$$

taking into account that the Hamiltonian H does not depend explicitly on time t and Eqs. (1.73) and (1.74).

In the case of a many-body system of N identical particles described by the Lagrangian (1.17), the canonical linear momenta are defined as

$$\mathbf{p}_i = \frac{\partial L}{\partial \dot{\mathbf{r}}_i} = m\dot{\mathbf{r}}_i , \tag{1.79}$$

while the Hamiltonian is

$$H = \sum_{i=1}^{N} \mathbf{p}_i \cdot \dot{\mathbf{r}}_i - L(\{\mathbf{r}_i\}, \{\dot{\mathbf{r}}_i\}) \tag{1.80}$$

or, explicitly,

$$H = \sum_{i=1}^{N} \left(\frac{p_i^2}{2m} + U(\mathbf{r}_i) \right) + \frac{1}{2} \sum_{\substack{i,j=1 \\ i \neq j}}^{N} V(\mathbf{r}_i, \mathbf{r}_j) . \tag{1.81}$$

Repeating the procedure illustrated above, one finds the Hamilton equations

$$\dot{\mathbf{r}}_i = \frac{\partial H}{\partial \mathbf{p}_i} , \tag{1.82}$$

$$\dot{\mathbf{p}}_i = -\frac{\partial H}{\partial \mathbf{r}_i} , \tag{1.83}$$

which imply

$$\dot{\mathbf{r}}_i = \frac{\mathbf{p}_i}{m} \tag{1.84}$$

$$\dot{\mathbf{p}}_i = -\frac{\partial}{\partial \mathbf{r}_i}\Big(U(\mathbf{r}_i) + \sum_{\substack{j=1 \\ j\neq i}}^{N} V(\mathbf{r}_i, \mathbf{r}_j) \Big) . \tag{1.85}$$

As expected, these $2N$ first-order differential equations are equivalent to the N second-order differential equations (1.20).

Let us consider the Schrödinger complex scalar field $\psi(\mathbf{r}, t)$ driven by the Lagrangian density (1.34). In this case the canonical momentum $\pi(\mathbf{r}, t)$ is a complex field defined as

$$\pi = \frac{\partial \mathscr{L}}{\partial(\partial_t \psi)} = \frac{i\hbar}{2}\psi^* , \tag{1.86}$$

while the canonical momentum $\bar{\pi}(\mathbf{r}, t)$ of $\psi^*(\mathbf{r}, t)$ is

$$\bar{\pi} = \frac{\partial \mathscr{L}}{\partial(\partial_t \psi^*)} = -\frac{i\hbar}{2}\psi . \tag{1.87}$$

Thus, we have $\bar{\pi}(\mathbf{r}, t) = \pi^*(\mathbf{r}, t)$. Notice that by using the asymmetric Lagrangian density (1.31), the canonical momenta are different, namely

$$\pi = \frac{\partial \mathscr{L}}{\partial(\partial_t \psi)} = i\hbar\psi^* \tag{1.88}$$

and

$$\bar{\pi} = 0. \tag{1.89}$$

The Hamiltonian density $\mathscr{H}$ is obtained performing the Legendre transform

$$\mathscr{H} = \pi \partial_t \psi + \bar{\pi} \partial_t \psi^* - \mathscr{L} . \tag{1.90}$$

In this way, in full generality we find

$$\mathscr{H} = \frac{\hbar^2}{2m}|\boldsymbol{\nabla}\psi|^2 + U\,|\psi|^2 + \frac{1}{2}\int d^3\mathbf{r}' |\psi(\mathbf{r}', t)|^2 V(\mathbf{r}, \mathbf{r}')|\psi(\mathbf{r}, t)|^2 . \tag{1.91}$$

Quite remarkably, the Hamiltonian density of the Schrödinger field does not depend on the canonical momenta $\pi(\mathbf{r}, t)$ and $\bar{\pi}(\mathbf{r}, t)$. For the sake of completeness, we underline that the Hamiltonian is given by

$$H = \int d^3\mathbf{r}\, \mathscr{H} \, . \tag{1.92}$$

Because of its relevance, we write it in full

$$\begin{aligned} H &= \int d^3\mathbf{r} \left(\frac{\hbar^2}{2m} |\boldsymbol{\nabla}\psi|^2 + U\, |\psi|^2 \right) \\ &\quad + \frac{1}{2} \int d^3\mathbf{r}\, d^3\mathbf{r}'\, |\psi(\mathbf{r}', t)|^2 V(\mathbf{r}, \mathbf{r}') |\psi(\mathbf{r}, t)|^2 \, . \end{aligned} \tag{1.93}$$

Sometimes, it is written as

$$\begin{aligned} H &= \int d^3\mathbf{r}\, \psi^* \left(-\frac{\hbar^2}{2m} \nabla^2 + U \right) \psi \\ &\quad + \frac{1}{2} \int d^3\mathbf{r}\, d^3\mathbf{r}'\, |\psi(\mathbf{r}', t)|^2 V(\mathbf{r}, \mathbf{r}') |\psi(\mathbf{r}, t)|^2 \, , \end{aligned} \tag{1.94}$$

which is, as previously stressed, an equivalent expression if the field $\psi(\mathbf{r}, t)$ vanishes at the boundaries of the spatial domain.

To conclude this section, we investigate the Hamiltonian formulation of the free electromagnetic field. The canonical momenta of the electromagnetic potentials are defined as

$$\Pi_0 = \frac{\partial \Phi}{\partial t} \, , \tag{1.95}$$

$$\boldsymbol{\Pi} = \frac{\partial \mathbf{A}}{\partial t} \, . \tag{1.96}$$

The Hamiltonian density is then given by

$$\mathscr{H} = \Phi\, \Pi_0 + \mathbf{A} \cdot \boldsymbol{P} - \mathscr{L} \tag{1.97}$$

where $\mathscr{L}$ is the Lagrangian density (1.46). After some calculations one finds the Hamiltonian

$$H = \int d^3\mathbf{r}\, \mathscr{H} = \int d^3\mathbf{r} \left\{ \frac{\epsilon_0}{2} (\boldsymbol{\nabla}\Phi + \partial_t \mathbf{A})^2 + \frac{1}{2\mu_0} (\boldsymbol{\nabla} \wedge \mathbf{A})^2 \right\} \tag{1.98}$$

which can also be rewritten in a more familiar way

$$H = \int d^3\mathbf{r} \left\{ \frac{\epsilon_0}{2}\mathbf{E}^2 + \frac{1}{2\mu_0}\mathbf{B}^2 \right\} . \tag{1.99}$$

Alternatively, working in the Coulomb gauge where $\Phi(\mathbf{r}, t) = 0$, the Hamiltonian can be expressed in terms of the vector potential $\mathbf{A}(\mathbf{r}, t)$ as

$$H = \int d^3\mathbf{r}\, \mathscr{H} = \int d^3\mathbf{r} \left\{ \frac{\epsilon_0}{2}(\partial_t \mathbf{A})^2 + \frac{1}{2\mu_0}(\boldsymbol{\nabla} \wedge \mathbf{A})^2 \right\} . \tag{1.100}$$

In conclusion we have seen that the two most fundamental formulations of classical mechanics of particles and fields are Lagrangian mechanics and Hamiltonian mechanics. Although the two formulations of mechanics may appear to differ greatly, but in reality, they are merely two viewpoints used to explain the same phenomena; in the end, they will ultimately produce the same outcomes and forecasts. The Hamiltonian approach is used to perform the canonical quantization, as we will see in the next sections. However, The Lagrangian formulation of quantum mechanics can also be used: this is known as the path integral quantum mechanics, that is the main topic of this book.

1.4 Canonical Quantization for Particles

According to the axiomatization of quantum mechanics developed by Paul Dirac and John von Neumann in the 1930s, the canonical quantization corresponds to promoting to quantum operators the coordinates and momenta of a classical Hamiltonian system by imposing very specific equal-time commutation rules. For a particle described by the Hamiltonian (1.67) we have

$$x(t) \rightarrow \hat{x}(t) , \tag{1.101}$$

$$p(t) \rightarrow \hat{p}(t) , \tag{1.102}$$

where the hat symbol is introduced to denote operators, and the coordinate operator $\hat{x}$ and the linear momentum operator $\hat{p}$ satisfy the following equal-time commutation rule

$$[\hat{x}(t), \hat{p}(t)] = i\hbar \, . \tag{1.103}$$

In this formula

$$[\hat{A}, \hat{B}] = \hat{A}\hat{B} - \hat{B}\hat{A} \tag{1.104}$$

is the so-called commutator of generic operators $\hat{A}$ and $\hat{B}$. As usual, $i = \sqrt{-1}$ is the imaginary unit and $\hbar$ is the reduced Planck constant. The Hamiltonian operator then reads

$$\hat{H} = H(\hat{x}, \hat{p}) = \frac{\hat{p}^2}{2m} + U(\hat{x}) \, . \tag{1.105}$$

The canonical quantization of one particle given by Eq. (1.101) and (1.102) is equivalent to the wave-like formulation of quantum mechanics due to Erwin Schrödinger by adopting the so-called coordinate representation of the quantization rules:

$$\hat{x} = x \, , \tag{1.106}$$

$$\hat{p} = -i\hbar \frac{\partial}{\partial x} \, . \tag{1.107}$$

Indeed, given a generic function $f(x)$ one finds immediately the commutation rule

$$\begin{aligned} (\hat{x}\,\hat{p} - \hat{p}\,\hat{x})\, f(x) &= -i\hbar\, x \frac{\partial}{\partial x} f(x) + i\hbar \frac{\partial}{\partial x}\, (x\, f(x)) \\ &= -i\hbar\, x \frac{\partial}{\partial x} f(x) + i\hbar\, f(x) + i\hbar\, x \frac{\partial}{\partial x} f(x) \\ &= i\hbar\, f(x) \, . \end{aligned} \tag{1.108}$$

Moreover, the coordinate representation of the quantization rules give immediately the quantum Hamiltonian operator in the coordinate representation

$$\hat{H} = -\frac{\hbar^2}{2m} \frac{\partial^2}{\partial x^2} + U(x) \tag{1.109}$$

from which one can write the time-dependent Schrödinger equation as

$$i\hbar \frac{\partial}{\partial t} \psi(x, t) = \hat{H} \psi(x, t) \, , \tag{1.110}$$

that is nothing else than Eq. (1.29). Using the Dirac notation of quantum states, the wavefunction can be written as

$$\psi(x) = \langle x|\psi\rangle \tag{1.111}$$

where $|x\rangle$ is the eigenstate of the coordinate operator $\hat{x}$ and $|\psi(t)\rangle$ is the quantum state belonging to the abstract Hilbert space, that is the domain of the Hamiltonian operator $\hat{H}$. In the abstract notation of Dirac, the time-dependent Schrödinger equation reads

$$i\hbar \frac{d}{dt}|\psi(t)\rangle = \hat{H}|\psi(t)\rangle \ , \tag{1.112}$$

where the symbol of the total derivative with respect to time t is usually adopted instead of the symbol of the partial derivative. The formal solution of this equation is

$$|\psi(t)\rangle = \hat{U}(t)\,|\psi(0)\rangle \tag{1.113}$$

with

$$\hat{U}(t) = e^{-i\hat{H}t/\hbar} \tag{1.114}$$

the unitary operator of the time evolution. It is indeed straightforward to recover Eq. (1.113) performing the time derivative of Eq. (1.112) taking into account Eq. (1.114).

For the many-body system described by the classical Hamiltonian (1.81) the quantization prescription

$$\mathbf{r}_j(t) \ \rightarrow \ \hat{\mathbf{r}}_j(t) \ , \tag{1.115}$$

$$\mathbf{p}_j(t) \ \rightarrow \ \hat{\mathbf{p}}_j(t) \ , \tag{1.116}$$

must the accompanied by the equal-time commutations rules

$$[\hat{x}_{jk}(t), \hat{p}_{js}(t)] = i\hbar\, \delta_{ks} \ , \tag{1.117}$$

where $\hat{\mathbf{r}}_j = (\hat{x}_{j1}, \hat{x}_{j2}, \hat{x}_{j3})$, $\hat{\mathbf{p}}_j = (\hat{p}_{j1}, \hat{p}_{j2}, \hat{p}_{j3})$, and δ_{ks} is the Kronecker delta. The Hamiltonian operator of the quantized system is then

$$\hat{H} = \sum_{j=1}^{N} \left(\frac{\hat{p}_j^2}{2m} + U(\hat{\mathbf{r}}_j) \right) + \frac{1}{2} \sum_{\substack{j,k=1 \\ j\neq k}}^{N} V(\hat{\mathbf{r}}_j, \hat{\mathbf{r}}_k) \ . \tag{1.118}$$

Finally, we observe that the abstract time-dependent Schrödinger equation of the many-body quantum state $|\Psi(t)\rangle$, given by

$$i\hbar\frac{\partial}{\partial t}|\Psi(t)\rangle = \hat{H}|\Psi(t)\rangle \ , \tag{1.119}$$

becomes

$$i\hbar\frac{\partial}{\partial t}\Psi = \left[\sum_{j=1}^{N}\left(-\frac{\hbar^2}{2m}\nabla_j^2 + U(\hat{\mathbf{r}}_j)\right) + \frac{1}{2}\sum_{\substack{j,k=1\\ j\neq k}}^{N} V(\hat{\mathbf{r}}_j,\hat{\mathbf{r}}_k)\right]\Psi \tag{1.120}$$

in the coordinate representation, where the many-body wavefunction $\psi = \psi(\mathbf{r}_1, ..., \mathbf{r}_N, t)$ corresponds to the scalar product $\langle \mathbf{r}_1...\mathbf{r}_N|\Psi(t)\rangle$, explicity

$$\Psi(\mathbf{r}_1, \mathbf{r}_2, ..., \mathbf{r}_{N-1}, \mathbf{r}_N, t) = \langle \mathbf{r}_1\mathbf{r}_2...\mathbf{r}_{N-1}\mathbf{r}_N|\Psi(t)\rangle \ , \tag{1.121}$$

where $|\mathbf{r}_1\mathbf{r}_2...\mathbf{r}_{N-1}\mathbf{r}_N\rangle = |\mathbf{r}_1\rangle|\mathbf{r}_2\rangle...|\mathbf{r}_{N-1}\rangle|\mathbf{r}_N\rangle$.

1.5 Canonical Quantization for Bosons

Inspired by the quantization of the electromagnetic field done by Dirac in 1927, the same year Pascual Jordan and Oscar Klein suggested that, in non-relativistic quantum mechanics, the quantum matter field is nothing but the single-particle Schrödinger field $\psi(\mathbf{r}, t)$ of quantum mechanics. In particular, the Hamiltonian operator of the quantum Schrödinger field $\hat{\psi}(\mathbf{r}, t)$ is given by

$$\begin{aligned}\hat{H} &= \int d^3\mathbf{r}\ \hat{\psi}^+(\mathbf{r})\left[-\frac{\hbar^2}{2m}\nabla^2 + U(\mathbf{r})\right]\hat{\psi}(\mathbf{r}) \\ &\quad + \frac{1}{2}\int d^3\mathbf{r}\ d^3\mathbf{r}'\ \hat{\psi}^+(\mathbf{r})\ \hat{\psi}^+(\mathbf{r}')\ V(\mathbf{r}-\mathbf{r}')\ \hat{\psi}(\mathbf{r}')\ \hat{\psi}(\mathbf{r}) \ .\end{aligned} \tag{1.122}$$

This quantum Hamiltonian can be directly obtained from the classical one, given by Eq. (1.94), by promoting the complex classical fields to quantum field operators:

$$\psi(\mathbf{r},t) \rightarrow \hat{\psi}(\mathbf{r},t) , \tag{1.123}$$

$$\psi^*(\mathbf{r},t) \rightarrow \hat{\psi}^+(\mathbf{r},t) , \tag{1.124}$$

$$\pi(\mathbf{r},t) \rightarrow \hat{\pi}(\mathbf{r},t) , \tag{1.125}$$

$$\pi^*(\mathbf{r},t) \rightarrow \hat{\pi}^+(\mathbf{r},t) . \tag{1.126}$$

It is crucial for these field operators the choice of the equal-time quantization rule. If $\hat{\psi}(\mathbf{r},t)$ describes an ensemble of identical bosonic particles (in a fixed spin state) one must choose the following bosonic equal-time commutation rule

$$[\hat{\psi}(\mathbf{r},t), \hat{\pi}(\mathbf{r}',t)] = i\hbar\, \delta(\mathbf{r}-\mathbf{r}') , \tag{1.127}$$

which are fully consistent with the commutation rules previously introduced for particles. According to Eq. (1.86) we have

$$\hat{\pi}(\mathbf{r},t) = \frac{i\hbar}{2}\, \hat{\psi}^+(\mathbf{r},t) \tag{1.128}$$

and consequently, in the nonrelativistic case, the equal-time commutation rule can also be written as

$$[\hat{\psi}(\mathbf{r},t), \hat{\psi}^+(\mathbf{r}',t)] = 2\delta(\mathbf{r}-\mathbf{r}') , \tag{1.129}$$

where $\hbar$ does not appear explicitly. On the other hand, adopting Eq. (1.88) we have

$$\hat{\pi}(\mathbf{r},t) = i\hbar\, \hat{\psi}^+(\mathbf{r},t) \tag{1.130}$$

and also

$$[\hat{\psi}(\mathbf{r},t), \hat{\psi}^+(\mathbf{r}',t)] = \delta(\mathbf{r}-\mathbf{r}') . \tag{1.131}$$

The formalism developed here is sometimes called second quantization because the Schördinger wavefunction $\psi(\mathbf{r},t)$ of (first) quantum mechanics is reinterpreted as the field operator $\hat{\psi}(\mathbf{r},t)$, that is the domain of quantum field theory (QFT). In this framework the equation of motion (1.35) becomes

$$i\hbar\frac{\partial}{\partial t}\hat{\psi}(\mathbf{r},t) = \left(-\frac{\hbar^2}{2m}\nabla^2 + U(\mathbf{r}) + \int d^3\mathbf{r}'\hat{\psi}^+(\mathbf{r},t)\,\hat{\psi}(\mathbf{r}',t)\,V(\mathbf{r},\mathbf{r}')\right) \times \hat{\psi}(\mathbf{r},t) , \tag{1.132}$$

that is the Heisenberg equation of the field operator $\hat{\psi}(\mathbf{r},t)$. Indeed, This equation can also be written as

$$i\hbar\frac{\partial}{\partial t}\hat{\psi}(\mathbf{r},t) = [\hat{\psi}(\mathbf{r},t),\hat{H}] \,. \tag{1.133}$$

The quantum field operator can be expanded as follows

$$\hat{\psi}(\mathbf{r},t) = \sum_{\alpha} \hat{c}_{\alpha}(t)\; \phi_{\alpha}(\mathbf{r}) \,, \tag{1.134}$$

where the $\phi_{\alpha}(\mathbf{r}) = \langle \mathbf{r}|\alpha\rangle$ are the eigenfunctions of the single-particle Hamiltonian

$$\hat{h} = -\frac{\hbar^2}{2m}\nabla^2 + U(\mathbf{r}) \tag{1.135}$$

such that

$$\hat{h}|\alpha\rangle = \epsilon_{\alpha}|\alpha\rangle \,, \tag{1.136}$$

and $\hat{c}_{\alpha}(t)$ and $\hat{c}_{\alpha}^{+}(t)$ are called ladder operators of the single-particle state $|\alpha\rangle$. From Eq. (1.129) one proves that the operators $\hat{c}_{\alpha}(t)$ and $\hat{c}_{\alpha}(t)^{+}$ satisfy the bosonic equal-time commutation relations

$$[\hat{c}_{\alpha}(t),\hat{c}_{\beta}^{+}(t)] = \delta_{\alpha,\beta} \,, \qquad [\hat{c}_{\alpha}(t),\hat{c}_{\beta}(t)] = [\hat{c}_{\alpha}^{+}(t),\hat{c}_{\beta}^{+}(t)] = 0 \,. \tag{1.137}$$

Working at fixed time t, it is relevant to introduce the single-mode number operators

$$\hat{N}_{\alpha} = \hat{c}_{\alpha}^{+}\hat{c}_{\alpha} \tag{1.138}$$

which counts the number of particles in the single-particle state $|\alpha\rangle$, i.e., they satisfy the following eigenvalue equation

$$\hat{N}_{\alpha}|\,...\,n_{\alpha}\,...\,\rangle = n_{\alpha}\;|\,...\,n_{\alpha}\,...\,\rangle \,, \tag{1.139}$$

where

$$|\,...\,n_{\alpha}\,...\,n_{\beta}\,...\,n_{\gamma}\,...\,\rangle \,, \tag{1.140}$$

is a generic multi-mode Fock state. The meaning of this multi-mode Fock state is that there are n_{α} particles in the single-particle state $|\alpha\rangle$, n_{β} particles in the single-particle state $|\beta\rangle$, n_{γ} particles in the single-particle state $|\gamma\rangle$, et cetera. The operators $\hat{c}_\alpha$ and $\hat{c}_\alpha^+$ are called annihilation and creation operators because they respectively destroy and create one particle in the single-particle state $|\alpha\rangle$, namely

$$\hat{c}_\alpha | \ldots n_\alpha \ldots \rangle = \sqrt{n_\alpha} \, | \ldots n_\alpha - 1 \ldots \rangle \, , \tag{1.141}$$
$$\hat{c}_\alpha^+ | \ldots n_\alpha \ldots \rangle = \sqrt{n_\alpha + 1} \, | \ldots n_\alpha + 1 \ldots \rangle \, . \tag{1.142}$$

Note that these properties follow directly from the commutation relations (1.137). Indeed the commutation rules (1.137) imply Eqs. (1.141) and (1.142) and, as expected for bosons, there is no restriction on the number of particles n_α which can occupy the single-particle state $|\alpha\rangle$. The vacuum state, where there are no particles, can be written as

$$|0\rangle = | \ldots 0 \ldots 0 \ldots 0 \ldots \rangle \, , \tag{1.143}$$

and

$$\hat{c}_\alpha |0\rangle = 0 \, , \tag{1.144}$$
$$\hat{c}_\alpha^+ |0\rangle = |1_\alpha\rangle = |\alpha\rangle \, , \tag{1.145}$$

where $|\alpha\rangle$ is such that

$$\langle \mathbf{r} | \alpha \rangle = \phi_\alpha(\mathbf{r}) \, . \tag{1.146}$$

Within the formalism of second quantization of field operators, the time-dependent one-body density operator is defined as

$$\hat{\rho}(\mathbf{r}, t) = \hat{\psi}^+(\mathbf{r}, t) \, \hat{\psi}(\mathbf{r}, t) \, , \tag{1.147}$$

and it is such that

$$\hat{N} = \int d^3\mathbf{r} \, \hat{\rho}(\mathbf{r}, t) \tag{1.148}$$

is the total number operator, which can be proven to be a constant of motion because

$$[\hat{N}, \hat{H}] = 0 \, . \tag{1.149}$$

By using the expansion (1.134) one finds immediately

$$\hat{\rho}(\mathbf{r}, t) = \sum_{\alpha,\beta} \hat{c}_\alpha^+(t) \hat{c}_\beta(t) \, \phi_\alpha^*(\mathbf{r}) \, \phi_\beta(\mathbf{r}) \tag{1.150}$$

and also

$$\hat{N} = \sum_{\alpha} \hat{c}_{\alpha}^{+}(t)\hat{c}_{\alpha}(t) = \sum_{\alpha} \hat{N}_{\alpha}(t) \, , \tag{1.151}$$

because we consider an orthonormal basis of single-particle wavefunctions $\phi_{\alpha}(\mathbf{r})$. Moreover, inserting Eq. (1.134) into Eq. (1.122) the quantum Hamiltonian can be rewritten as

$$\hat{H} = \sum_{\alpha} \epsilon_{\alpha} \, \hat{c}_{\alpha}^{+}\hat{c}_{\alpha} + \frac{1}{2} \sum_{\alpha\beta\gamma\delta} V_{\alpha\beta\gamma\delta} \, \hat{c}_{\alpha}^{+}\hat{c}_{\beta}^{+}\hat{c}_{\delta}\hat{c}_{\gamma} \, , \tag{1.152}$$

where

$$V_{\alpha\beta\delta\gamma} = \int d^3\mathbf{r} \, d^3\mathbf{r}' \; \phi_{\alpha}^{*}(\mathbf{r}) \; \phi_{\beta}^{*}(\mathbf{r}') \; V(\mathbf{r}-\mathbf{r}') \; \phi_{\delta}(\mathbf{r}') \; \phi_{\gamma}(\mathbf{r}) \, . \tag{1.153}$$

This representation of the QFT Hamiltonian operator is often used in the quantum many-body theory of condensed matter.

1.6 Canonical Quantization for Fermions

The Schrödinger field operator $\hat{\psi}(\mathbf{r}, t)$ can be used to describe also an ensemble of identical fermionic particles (in a fixed spin state). In this case the Hamiltonian (1.122), the dynamical equation (1.132), and the expansion (1.134) do not change. However, to obtain fermionic properties, it is necessary to impose the following fermionic equal-time anti-commutation rule

$$\{\hat{\psi}(\mathbf{r}, t), \hat{\pi}(\mathbf{r}', t)\} = i\hbar \, \delta(\mathbf{r} - \mathbf{r}') \tag{1.154}$$

or, equivalently

$$\{\hat{\psi}(\mathbf{r}, t), \hat{\psi}^{+}(\mathbf{r}', t)\} = \delta(\mathbf{r} - \mathbf{r}') \, , \tag{1.155}$$

where

$$\{\hat{A}, \hat{B}\} = \hat{A}\hat{B} + \hat{B}\hat{A} \tag{1.156}$$

are the anti-commutation brackets. In addition one must also impose

$$\{\hat{\psi}(\mathbf{r}, t), \hat{\psi}(\mathbf{r}', t')\} = \{\hat{\psi}^{+}(\mathbf{r}, t), \hat{\psi}^{+}(\mathbf{r}', t')\} = 0 \, . \tag{1.157}$$

As a consequence, there are fermionic anti-commutation rules for the operators $\hat{c}_\alpha$ and $\hat{c}_\alpha^+$, i.e.

$$\{\hat{c}_\alpha, \hat{c}_\beta^+\} = \delta_{\alpha\beta} , \qquad \{\hat{c}_\alpha, \hat{c}_\beta\} = \{\hat{c}_\alpha^+, \hat{c}_\beta^+\} = 0 . \tag{1.158}$$

Moreover, it follows that

$$(\hat{c}_\alpha^+)^2 = 0 . \tag{1.159}$$

Notice that Eqs. (1.141) and (1.142) formally work also for fermions but with the constraints that $n_\alpha = 0$ or $n_\alpha = 1$ for any single-particle state $|\alpha\rangle$. In fact, for fermions

$$\hat{N}_\alpha^2 = \left(\hat{c}_\alpha^+ \hat{c}_\alpha\right)^2 = \hat{c}_\alpha^+ \hat{c}_\alpha \hat{c}_\alpha^+ \hat{c}_\alpha = \hat{c}_\alpha^+ \left(1 - \hat{c}_\alpha^+ \hat{c}_\alpha\right) \hat{c}_\alpha = \hat{c}_\alpha^+ \hat{c}_\alpha = \hat{N}_\alpha , \tag{1.160}$$

which implies that the eigenvalues of $\hat{N}_\alpha$ can be only 0 and 1. For a fermionic two-particle state, we have

$$|1_\alpha\, 1_\beta\rangle = \hat{c}_\alpha^+ \hat{c}_\beta^+ |0\rangle = -\hat{c}_\beta^+ \hat{c}_\alpha^+ |0\rangle = -|1_\beta\, 1_\alpha\rangle , \tag{1.161}$$

i.e., the state is anti-symmetric under interchange of particle labels, and also

$$|2_\alpha\rangle = (\hat{c}_\alpha^+)^2 |0\rangle = 0 , \tag{1.162}$$

expressing the Pauli principle that two particles cannot be created in the same single-particle state. For fermions the field operator $\hat{\psi}(\mathbf{r}, t)$ is called fermionic field operator.

In the case of electrons, which are fermions with two possible spin states $\sigma = \uparrow, \downarrow$, the spin degree of freedom can be inserted as a subindex in the field operator, i.e., $\psi_\sigma(\mathbf{r}, t)$. Consequently, the qauntum-field-theory Hamiltonian of interacting electrons reads

$$\begin{aligned} \hat{H} &= \sum_{\sigma=\uparrow,\downarrow} \int d^3\mathbf{r}\ \hat{\psi}_\sigma^+(\mathbf{r}) \left[-\frac{\hbar^2}{2m}\nabla^2 + U(\mathbf{r})\right] \hat{\psi}_\sigma(\mathbf{r}) \\ &+ \frac{1}{2} \sum_{\sigma,\sigma'=\uparrow,\downarrow} \int d^3\mathbf{r}\ d^3\mathbf{r}'\ \hat{\psi}_\sigma^+(\mathbf{r})\ \hat{\psi}_{\sigma'}^+(\mathbf{r}')\ V(\mathbf{r}-\mathbf{r}')\ \hat{\psi}_{\sigma'}(\mathbf{r}')\ \hat{\psi}_\sigma(\mathbf{r}) \end{aligned} \tag{1.163}$$

and the expansion of the electronic field becomes

$$\hat{\psi}_\sigma(\mathbf{r}, t) = \sum_\alpha \hat{c}_{\alpha,\sigma}(t)\, \phi_\alpha(\mathbf{r}, t) . \tag{1.164}$$

This means that the ladder operator $\hat{c}_{\alpha,\sigma}$ annihilates an electron of spin σ from the single-particle quantum state $|\alpha\rangle$ while the operator $\hat{c}^+_{\alpha,\sigma}$ creates an electron of spin σ in the single-partial quantum state $|\alpha\rangle$.

1.7 Canonical Quantization for Photons

Photons are relativistic massless particles of light, and they are characterized by the following dispersion relation in vacuum

$$\omega_k = c\, k \,, \tag{1.165}$$

where ω_k is the angular frequency, $c = 3\times10^8$ meters/seconds is the speed of light, $\mathbf{k}$ is the wavevector, and $k = |\mathbf{k}|$ is the wavenumber. In general, a photon has two possible polarization states $s = 1, 2$. The QFT of photons in free space is described by a Hamiltonian that is a simplified version of Eq. (1.152), namely

$$\hat{H} = \sum_s \sum_{\mathbf{k}} \hbar\omega_k\; \hat{a}^+_{\mathbf{k}s}\hat{a}_{\mathbf{k}s} = \sum_s \sum_{\mathbf{k}} \hbar\omega_k\; \hat{N}_{\mathbf{k}s} \,, \tag{1.166}$$

where $\hat{a}_{\mathbf{k}s}$ and $\hat{a}^+_{\mathbf{k}s}$ are the ladder operators that annihilate and create a photon with wavevector $\mathbf{k}$ and polarization s. For instance,

$$|\mathbf{k}s\rangle = \hat{a}^+_{\mathbf{k}s}|0\rangle \tag{1.167}$$

is the quantum Fock state of a single photon with wavevector $\mathbf{k}$ and polarization s. Instead,

$$|\mathbf{k}s, \mathbf{k}'s'\rangle = |\mathbf{k}s\rangle|\mathbf{k}'s'\rangle = \hat{a}^+_{\mathbf{k}s}\hat{a}^+_{\mathbf{k}'s'}|0\rangle \tag{1.168}$$

is the quantum Fock state of a two-photon with wavevectors $\mathbf{k}$ and $\mathbf{k}'$ and polarizations s and s'. The relativistic and massless nature of photons is encoded in the dispersion relation, but the structure of the Hamiltonian in second quantization is similar to that one of nonrelativistic and massive bosons. As usual, $\hat{N}_{\mathbf{k}s} = \hat{a}^+_{\mathbf{k}s}\hat{a}_{\mathbf{k}s}$ is the number operator for the mode with wavevector $\mathbf{k}$ and polarization s.

Much more complicated is the quantum field theory of photons in a medium. In fact, in general the dispersion relation will be

$$\omega_k = v_0 + v_1\, k + v_2\, k^2 + v_3\, k^3 + ... \,, \tag{1.169}$$

where the parameters v_0, v_1, v_2, v_3, ... can depend on the number of photons. The corresponding Hamiltonian could be

$$\begin{aligned}\hat{H} \;&=\; \sum_{s=\pm 1}\sum_{\mathbf{k}} \hbar\Big(v_0(\hat{N}_{\mathbf{k}s}) + v_1(\hat{N}_{\mathbf{k}s})\, k + v_2(\hat{N}_{\mathbf{k}s})\, k^2 \\ &\quad + v_3(\hat{N}_{\mathbf{k}s})\, k^3 + ...\Big)\, \hat{N}_{\mathbf{k}s}\,. \end{aligned} \tag{1.170}$$

Just to make an explicit example, for photons in nonlinear Kerr media only the speed parameter v_1 can be taken differently from zero. It can be approximated as

$$v_1 = v_1^{(0)} + v_1^{(1)}\, \hat{N}_{\mathbf{k}s} \tag{1.171}$$

and the Hamiltonian reads

$$\hat{H} = \sum_{s=\pm 1}\sum_{\mathbf{k}} \left[\hbar v_1^{(0)} k\, \hat{N}_{\mathbf{k}s} + \hbar v_1^{(1)} k\, \hat{N}_{\mathbf{k}s}\, \hat{N}_{\mathbf{k}s}\right]\,. \tag{1.172}$$

In the Coulomb gauge, also called the transverse gauge, the derivation of the quantum Hamiltonian (1.166) from the classical Hamiltonian (1.100) requires promoting the vector potential $\mathbf{A} = (A_1, A_2, A_3)$ and its conjugate momentum $\mathbf{\Pi} = (\Pi_1, \Pi_2, \Pi_3)$ to operators

$$\mathbf{A}(\mathbf{r}, t) \to \hat{\mathbf{A}}(\mathbf{r}, t) \tag{1.173}$$

$$\mathbf{\Pi}(\mathbf{r}, t) \to \hat{\mathbf{\Pi}}(\mathbf{r}, t) \tag{1.174}$$

imposing these equal-time commutation rules

$$[\hat{A}_j(\mathbf{r}, t), \hat{\Pi}_k(\mathbf{r}', t)] = i\hbar\, \hat{\delta}^{\perp}_{jk}\, \delta^{(3)}(\mathbf{r} - \mathbf{r}') \tag{1.175}$$

with

$$\hat{\delta}^{\perp}_{jk} = \delta_{jk} - \frac{\partial_j \partial_k}{\nabla^2} \tag{1.176}$$

the transverse Kronecker delta, which is an operator ensuring that the functions acting on it are projected into their transverse components. The vector potential operator $\hat{\mathbf{A}}(\mathbf{r}, t)$ can be expressed as a Fourier series of monochromatic plane waves

$$\hat{\mathbf{A}}(\mathbf{r}, t) = \sum_{\mathbf{k}} \sum_{s} \sqrt{\frac{\hbar}{2\varepsilon_0 \omega_k}} \left[\hat{a}_{\mathbf{k}s}(t) \frac{e^{i\mathbf{k}\cdot\mathbf{r}}}{\sqrt{V}} + \hat{a}^{+}_{\mathbf{k}s}(t) \frac{e^{-i\mathbf{k}\cdot\mathbf{r}}}{\sqrt{V}} \right] \mathbf{u}_{\mathbf{k}s} , \tag{1.177}$$

where $\hat{a}_{\mathbf{k}s}(t)$ and $\hat{a}^{+}_{\mathbf{k}s}(t)$ are adimensional operator coefficients of the expansion, the complex plane waves $e^{i\mathbf{k}\cdot\mathbf{r}}/\sqrt{V}$ is normalized in a volume V are the basis functions of the expansion, and $\mathbf{u}_{\mathbf{k}1}$ and $\mathbf{u}_{\mathbf{k}2}$ are two mutually orthogonal real unit vectors of polarization which are also orthogonal to $\mathbf{k}$ (transverse polarization vectors). From the commutation rules (1.176) one gets

$$[\hat{a}_{\mathbf{k}s}(t), \hat{a}^{+}_{\mathbf{k}'s'}(t)] = \delta_{ss'}\, \delta_{\mathbf{k}\mathbf{k}'} \tag{1.178}$$

veryfing that $\hat{a}_{\mathbf{k}s}(t)$ and $\hat{a}^{+}_{\mathbf{k}s}(t)$ are bosonic ladder operators. Inserting Eq. (1.177) into Eq. (1.100), one finally obtains the quantum Hamiltonian (1.166).

1.8 Quantum Optics and Coherent States

In the context of quantum optics Roy Glauber and George Sudarshan introduced in 1960s the coherent state $|c\rangle$ as the eigenstate of the annihilation operator $\hat{c}$ with complex eigenvalue $c = |c|e^{i\theta}$, namely

$$\hat{c}|c\rangle = c|c\rangle , \tag{1.179}$$

with

$$\langle c|c\rangle = 1 . \tag{1.180}$$

Here $\hat{c}$ and $\hat{c}^{+}$ are generic ladder operators satisfying the commutation rule $[\hat{c}, \hat{c}^{+}] = 1$. The coherent state is very useful in optics because, as proven experimentally by Fortunato Tito Arecchi in 1965, the radiation field of a well-stabilized laser device operating in a single mode is accurately described by a coherent state $|c\rangle$.

It is important to stress that $|c\rangle$ does not have a fixed number of particles, i.e., it is not an eigenstate of the number operator $\hat{N} = \hat{c}^+\hat{c}$, and it is not difficult to show that $|c\rangle$ can be expanded in terms of number (Fock) states $|n\rangle$ as follows

$$|c\rangle = e^{-|c|^2/2} \sum_{n=0}^{\infty} \frac{c^n}{\sqrt{n!}} |n\rangle \ . \tag{1.181}$$

Let us prove this formula. The coherent state $|c\rangle$ can be expanded as

$$|c\rangle = \sum_{n=0}^{\infty} a_n |n\rangle \ , \tag{1.182}$$

where a_n are the complex coefficients to be determined. Then we have

$$\begin{aligned} \hat{c}|c\rangle &= \hat{c} \sum_{n=0}^{\infty} a_n |n\rangle = \sum_{n=0}^{\infty} a_n \hat{c} |n\rangle = \sum_{n=1}^{\infty} a_n \sqrt{n} |n-1\rangle \\ &= \sum_{n=0}^{\infty} a_{n+1} \sqrt{n+1} |n\rangle \end{aligned} \tag{1.183}$$

but also

$$c|c\rangle = c \sum_{n=0}^{\infty} a_n |n\rangle = \sum_{n=0}^{\infty} c\, a_n |n\rangle \ . \tag{1.184}$$

Since, Eq. (1.179) holds by definition, it follows that

$$a_{n+1}\sqrt{n+1} = c\, a_n \tag{1.185}$$

from which

$$a_{n+1} = \frac{c}{\sqrt{n+1}} a_n \ , \tag{1.186}$$

namely

$$a_1 = \frac{c}{\sqrt{1}} a_0 \ , \ a_2 = \frac{c^2}{\sqrt{2!}} a_0 \ , \ a_3 = \frac{c^3}{\sqrt{3!}} a_0 \ , \ ... \tag{1.187}$$

and in general

$$a_n = \frac{c^n}{\sqrt{n!}} a_0 \ . \tag{1.188}$$

Summarizing

$$|c\rangle = \sum_{n=0}^{\infty} a_0 \frac{c^n}{\sqrt{n!}} |n\rangle , \tag{1.189}$$

where the parameter a_0 is fixed by the normalization

$$1 = \langle c|c\rangle = |a_0|^2 \sum_{n=0}^{\infty} \frac{c^{2n}}{n!} = |a_0|^2 \, e^{|c|^2} , \tag{1.190}$$

and consequently

$$a_0 = e^{-|c|^2/2} , \tag{1.191}$$

up to a constant phase factor.

Taking into account that a Fock state $|n\rangle$ can be written in terms of the vacuum state $|0\rangle$ as follows

$$|n\rangle = \frac{1}{\sqrt{n!}} \hat{c}^+ |0\rangle , \tag{1.192}$$

inserting this expression into Eq. (1.181) we get

$$|c\rangle = e^{-|c|^2/2} \, e^{c\hat{c}^+} |0\rangle \tag{1.193}$$

because

$$e^{c\hat{c}^+} = \sum_{n=0}^{\infty} \frac{(c\hat{c}^+)^n}{n!} . \tag{1.194}$$

From Eq. (1.179) one immediately finds

$$\bar{N} = \langle c|\hat{N}|c\rangle = |c|^2 , \tag{1.195}$$

and it is natural to set

$$c = \sqrt{\bar{N}} \, e^{i\theta} , \tag{1.196}$$

where $\bar{N}$ is the average number of photons in the coherent state, while θ is the phase of the coherent state. For the sake of completeness, we observe that

$$\langle c|\hat{N}^2|c\rangle = |c|^2 + |c|^4 = \bar{N} + \bar{N}^2 \tag{1.197}$$

and consequently

$$\langle c|\hat{N}^2|c\rangle - \langle c|\hat{N}|c\rangle^2 = \bar{N} \ , \tag{1.198}$$

while

$$\langle n|\hat{N}^2|n\rangle = n^2 \tag{1.199}$$

and this implies

$$\langle n|\hat{N}^2|n\rangle - \langle n|\hat{N}|n\rangle^2 = 0 \ . \tag{1.200}$$

Quite important is the probability of finding the coherent state $|c\rangle$ in the Fock state $|n\rangle$. This probability is given by

$$p = |\langle n|c\rangle|^2 \ . \tag{1.201}$$

Because

$$\begin{aligned} \langle n|c\rangle &= \langle n|\sum_{m=0}^{\infty} e^{-|c|^2/2}\frac{c^m}{\sqrt{m!}}|m\rangle = \sum_{m=0}^{\infty} e^{-|c|^2/2}\frac{c^m}{\sqrt{m!}}\langle n|m\rangle \\ &= e^{-|c|^2/2}\frac{c^n}{\sqrt{n!}} \ , \end{aligned} \tag{1.202}$$

we get

$$p = c^{-|c|^2}\ \frac{|c|^{2n}}{n!} \ , \tag{1.203}$$

which is the familiar Poisson distribution. This Poission distribution was experimentally found by Tito Arecchi at the University of Milan in 1965, investigating a single-mode laser. Quite generally, the coherent state gives photocount statistics that are in accord with laser experiments and has coherence properties similar to those of a classical field, which are crucial for interference effects.

The probability of finding the coherent state $|c\rangle$ in the coherent state $|c'\rangle$ is defined as

$$p = |\langle c'|c\rangle|^2 \ . \tag{1.204}$$

Taking into account

$$\begin{aligned}\langle c'|c\rangle &= \langle c'|\sum_{m=0}^{\infty} e^{-|c|^2/2}\frac{c^m}{\sqrt{m!}}|m\rangle \\ &= \sum_{n=0}^{\infty}\sum_{m=0}^{\infty} e^{-|c'|^2/2}e^{-|c|^2/2}\frac{((c')^*)^n}{\sqrt{n!}}\frac{c^m}{\sqrt{m!}}\langle n|m\rangle \\ &= e^{-(|c|^2+|c'|^2)/2}\sum_{n=0}^{\infty}\frac{((c')^*c)^n}{n!} = e^{-(|c|^2+|c'|^2)/2}\, e^{(c')^*c} \\ &= e^{-(|c|^2+|c'|^2-2c(c')^*)/2} \end{aligned} \tag{1.205}$$

we find

$$p = e^{-|c-c'|^2} \,. \tag{1.206}$$

This means that two generic coherent states $|c\rangle$ and $|c'\rangle$ are never orthogonal.

A relevant property of coherent states $|c\rangle$ is the spectral resolution of the identity

$$\hat{1} = \int_{\mathbb{C}} \frac{dc^*\, dc}{2\pi i}\, |c\rangle\langle c| \,. \tag{1.207}$$

This formula can be proved as follows: given two generic Fock states $|n\rangle$ and $|m\rangle$, we have

$$\begin{aligned}\langle n|\left(\int_{\mathbb{C}} \frac{dc^*\, dc}{2\pi i}\, |c\rangle\langle c|\right)|m\rangle &= \int_{\mathbb{C}} \frac{dc^*\, dc}{2\pi i}\langle n|c\rangle\langle c|m\rangle \\ &= \int_{\mathbb{C}} \frac{dc^*\, dc}{2\pi i} e^{-|c|^2}\frac{c^n(c^*)^m}{\sqrt{n!m!}} \\ &= \int_0^{+\infty} d\rho\, \rho \int_0^{2\pi}\frac{d\theta}{\pi}\, e^{-\rho^2}\, \rho^{n+m}\, \frac{e^{i(n-m)\theta}}{\sqrt{n!m!}} \end{aligned} \tag{1.208}$$

setting $c = \rho\, e^{i\theta}$. Then we obtain

$$\begin{aligned} &\frac{2}{\sqrt{n!m!}}\int_0^{+\infty} d\rho\, \rho^{n+m+1}\, e^{-\rho^2}\int_0^{2\pi}\frac{d\theta}{2\pi}\, e^{i(n-m)\theta} \\ &\quad = \frac{2}{\sqrt{n!m!}}\int_0^{+\infty} d\rho\, \rho^{n+m+1}\, e^{-\rho^2}\, \delta_{n,m} \\ &\quad = \frac{2}{n!}\int_0^{+\infty} d\rho\, \rho^{2n+1}\, e^{-\rho^2}\delta_{n,m} = \delta_{n,m} \,. \end{aligned} \tag{1.209}$$

Thus

$$\langle n| \left(\int_{\mathbb{C}} \frac{dc^* \, dc}{2\pi i} \, |c\rangle\langle c| \right) |m\rangle = \delta_{n,m} = \langle n|m\rangle \tag{1.210}$$

and this implies that Eq. (1.207) is true.

1.9 Coherent States for Bosonic Fields

The coherent states are a formidable mathematical tool for laser physics. However, they are also very important for matter fields. Indeed, they are the theoretical core of all the macroscopic quantum phenomena. Among them, in addition to the laser light, the Bose-Einstein condensation with dilute and ultracold bosonic gases made of alkali-metal atoms, the superfluidity of quantum liquids (^{4}He, ^{3}He, and neutron stars), and the superconductivity of metals, alloy, cuprates, and other composite solid-state materials.

As an example, let us consider the relevant case of a dilute gas of ^{87}Rb atoms at the temperature of 100 nK in the lowest single-particle state under a confining magnetic or optical potential $U(\mathbf{r})$. Eric Cornell, Carl Wieman and Wolfgang Ketterle got the Nobel prize in Physics (2001) for seminal experiments on the Bose-Einstein condensation with ultracold atoms. The atomic gas can be so dilute that, under the effect of an external magnetic field which tunes the inter-atomic strength, the system becomes effectively non-interacting. Under these conditions, one has a pure Bose-Einstein condensate made of non-interacting bosons described by the Hamiltonian

$$\hat{H} = \int d^3\mathbf{r} \, \hat{\psi}^+(\mathbf{r}) \left[-\frac{\hbar^2}{2m}\nabla^2 + U(\mathbf{r}) \right] \hat{\psi}(\mathbf{r}) = \sum_\alpha \epsilon_\alpha \, \hat{c}_\alpha^+ \hat{c}_\alpha \, . \tag{1.211}$$

If there are exactly N bosons in the lowest single-particle state of energy ϵ_0, the ground state of the system is a Fock state given by

$$|F_N\rangle = \frac{1}{\sqrt{N!}} \left(\hat{c}_0^+ \right)^N |0\rangle = |N, 0, 0, ... \rangle \, . \tag{1.212}$$

It is then straightforward to show that

$$\langle F_N|\hat{\psi}(\mathbf{r})|F_N\rangle = 0 \ , \tag{1.213}$$

while the expectation value of $\hat{\psi}^+(\mathbf{r})\hat{\psi}(\mathbf{r})$ is given by

$$\langle F_N|\hat{\psi}^+(\mathbf{r})\hat{\psi}(\mathbf{r})|F_N\rangle = N\,|\psi_0(\mathbf{r})|^2 \ . \tag{1.214}$$

These results are the exact analog of what we have seen for the light field in Chapter 2. In the spirit of quantum optics one can suppose that the number of massive bosons in the matter field is not fixed, in other words that the system is not in a Fock state. We then introduce the coherent state $|c_0\rangle$, such that

$$\hat{c}_0|c_0\rangle = c_0|c_0\rangle \ , \tag{1.215}$$

with

$$\langle c_0|c_0\rangle = 1 \ . \tag{1.216}$$

The coherent state $|c_0\rangle$ is thus the eigenstate of the annihilation operator $\hat{c}_0$ with complex eigenvalue $c_0 = |c_0|e^{i\theta_0}$. $|c_0\rangle$ does not have a fixed number of bosons, i.e., it is not an eigenstate of the number operator $\hat{N}$ nor of $\hat{N}_0$, and it is not difficult to show that $|c_0\rangle$ can be expanded in terms of number (Fock) states $|N, 0, 0, ...\rangle$ as follows

$$|c_0\rangle = e^{-|c_0|^2/2}\sum_{N=0}^{\infty}\frac{c_0^N}{\sqrt{N!}}|N, 0, 0, ...\rangle \ . \tag{1.217}$$

From Eq. (1.215) one immediately finds

$$\bar{N} = \langle c_0|\hat{N}|c_0\rangle = |c_0|^2 \ , \tag{1.218}$$

and it is natural to set

$$c_0 = \sqrt{\bar{N}}\,e^{i\theta_0} \ , \tag{1.219}$$

where $\bar{N}$ is the average number of bosons in the coherent state, while θ_0 is the phase of the coherent state. The expectation value of the matter field $\hat{\psi}(\mathbf{r})$ in the coherent state $|c_0\rangle$ reads

$$\langle c_0|\hat{\psi}(\mathbf{r})|c_0\rangle = \sqrt{\bar{N}}\,e^{i\theta_0}\,\phi_0(\mathbf{r}) \ , \tag{1.220}$$

with $\omega_0 = \epsilon_0/\hbar$, while the expectation value of $\hat{\psi}^+(\mathbf{r})\hat{\psi}(\mathbf{r})$ is given by

$$\langle c_0|\hat{\psi}^+(\mathbf{r})\hat{\psi}(\mathbf{r})|c_0\rangle = \bar{N}\,|\phi_0(\mathbf{r})|^2 \,. \tag{1.221}$$

In full generality, the classical analog of the bosonic quantum field operator

$$\hat{\psi}(\mathbf{r}) = \sum_\alpha \phi_\alpha(\mathbf{r})\;\hat{c}_\alpha \tag{1.222}$$

is the complex classical field

$$\psi(\mathbf{r}) = \sum_\alpha \phi_\alpha(\mathbf{r})\;c_\alpha \tag{1.223}$$

such that

$$\hat{\psi}(\mathbf{r})|\psi\rangle = \psi(\mathbf{r})|\psi\rangle \,, \tag{1.224}$$

where

$$|\psi\rangle = \prod_\alpha |c_\alpha\rangle \tag{1.225}$$

is the bosonic coherent state of the system, $|c_\alpha\rangle$ is the coherent state of the bosonic operator $\hat{c}_\alpha$, and c_α is its complex eigenvalue. Finally, we underline that the adjoint bosonic field

$$\psi^+(\mathbf{r}) = \sum_\alpha \phi^*_\alpha(\mathbf{r})\;\hat{c}^+_\alpha \tag{1.226}$$

has the corresponding classical analog

$$\psi^+(\mathbf{r}) = \sum_\alpha \phi^*_\alpha(\mathbf{r})\;c^*_\alpha \tag{1.227}$$

where c^*_α is the complex conjugate of c_α.

1.10 Coherent States for Fermionic Fields

It is possible to introduce coherent states also for the fermionic field. For simplicity, let us start with a single mode of the fermionic field described by the anti-commuting fermionic operators $\hat{c}$ and $\hat{c}^+$, such that

$$\hat{c}\hat{c}^+ + \hat{c}^+\hat{c} = 1 \, , \qquad \hat{c}^2 = (\hat{c}^+)^2 = 0 \, . \tag{1.228}$$

The fermionic coherent state $|c\rangle$ is defined as the eigenstate of the fermionic annihilation operator $\hat{c}$, namely

$$\hat{c}|c\rangle = c|c\rangle \, , \tag{1.229}$$

where c is the corresponding eigenvalue, but also where $\bar{c}$ is such that

$$\langle c|\hat{c}^+ = \bar{c}\langle c| \tag{1.230}$$

with $\bar{c}$ a sort of complex conjugate of c. It is then reasonable to assume that these numbers c and $\bar{c}$ satisfy the following relationships

$$c^2 = \bar{c}^2 = 0 \, . \tag{1.231}$$

Thus, both c and $\bar{c}$ anti-commute with themselves. Obviously, c and $\bar{c}$ are not complex numbers but very special objects, invented by Hermann Günther Grassmann in 1844. They are called complex Grassmann numbers, namely elements of the Grassmann linear algebra characterized by the independent basis elements 1, c, $\bar{c}$, with 1 the identity (neutral) element. Usually $\bar{c}$ is called conjugate Grassmann number of the Grassmann number c. The two Grassmann numbers c and $\bar{c}$ are, by definition, independent and also anti-commuting with each other, i.e.

$$c\,\bar{c} = -\bar{c}\,c \, . \tag{1.232}$$

By definition, the Grassmann numbers commute with the complex numbers and with bosonic ladder operators. Instead, the Grassmann numbers anti-commute with fermionic ladder operators. For instance,

$$c\,\hat{c} = -\hat{c}\,c \, , \qquad c\,\hat{c}^+ = -\hat{c}^+\,c \, . \tag{1.233}$$

The most general function on this Grassmann algebra is given by

$$f(\bar{c}, c) = \alpha + \beta\, c + \gamma\, \bar{c} + \delta\, \bar{c}\, c \, , \tag{1.234}$$

where α, β, γ, δ are complex numbers. In fact, the function $f(\bar{c}, c)$ does not have higher powers of c, $\bar{c}$, and $\bar{c}\,c$ because they are identically zero.

The differentiation with respect to the Grassmann variable c is defined as

$$\frac{\partial}{\partial c} f(\bar{c}, c) = \beta - \delta\, \bar{c}\,, \tag{1.235}$$

where the minus sign occurs because one needs to permute $\bar{c}$ and c before differentiation. This is called left differentiation. Clearly, one has also

$$\frac{\partial}{\partial \bar{c}} f(\bar{c}, c) = \alpha + \delta\, c \tag{1.236}$$

and

$$\frac{\partial^2}{\partial \bar{c} \partial c} f(\bar{c}, c) = -\frac{\partial^2}{\partial c \partial \bar{c}} f(\bar{c}, c) = -\delta\,. \tag{1.237}$$

The integration over Grassmann numbers is much more trickier. The mathematical theory of this kind of Grassmann integral (now also called Berezin integral) with anti-commuting Grassmann variables was invented by David John Candlin in 1956 and developed by Felix Berezin in 1966. The Grassmann-Berezin integral is mainly used for the path integral formulation of quantum field theory with fermions. These integrals are not integrals in the Lebesgue sense. They are called integrals because they have some basic properties of integrals: the linearity

$$\int dc\; (\alpha f(c) + \beta g(c)) = \alpha \int dc\, f(c) + \beta \int dc\, g(c)\,, \tag{1.238}$$

a sort of Torricelli-Barrow formula

$$\int dc \frac{df}{dc} = 0 \tag{1.239}$$

as if $f(c)$ vanishes at infinity, the identities

$$\int dc = 0\,, \tag{1.240}$$

$$\int dc\, c = 1\,, \tag{1.241}$$

and the shift-invariance property

$$\int dc\, f(c+\xi) = \int dc\, f(c) \tag{1.242}$$

with ξ another Grassmann number, which is inspired by the similar property

$$\int_{-\infty}^{+\infty} dx\, f(x+y) = \int_{-\infty}^{+\infty} dx\, f(x) \tag{1.243}$$

holding for the improper integral of real variables x and y.

In practice, the Grassmann-Berezin integral is equivalent to Grassmann differentiation, i.e.

$$\int dc = \frac{\partial}{\partial c}\,, \quad \int d\bar{c} = \frac{\partial}{\partial \bar{c}}\,, \quad \int d\bar{c}\, dc = \frac{\partial^2}{\partial \bar{c} \partial c}\,, \quad \int dc\, d\bar{c} = \frac{\partial^2}{\partial c \partial \bar{c}}\,. \tag{1.244}$$

For this reason, the change of variable

$$c = A\, c' \tag{1.245}$$

corresponds to a quite unusual change in the symbol of the differential

$$dc = dc'\, A^{-1}\,. \tag{1.246}$$

This unusual formula is, however, consistent with the requirement that Eq. (1.241) holds for any Grassmann number:

$$1 = \int dc\, c = \int dc'\, A^{-1}\, A\, c' = \int dc'\, c'\,. \tag{1.247}$$

Due to the anti-commutation properties of Grassmann numbers, the fermionic coherent state $|c\rangle$ can be written in terms of the Fock states $|0\rangle$ and $|1\rangle$ as

$$|c\rangle = e^{-c\hat{c}^+}|0\rangle = \left(1 - c\hat{c}^+\right)|0\rangle = |0\rangle - c|1\rangle\,. \tag{1.248}$$

In fact, we have

$$\hat{c}\,(|0\rangle - c|1\rangle) = c\hat{c}|1\rangle = c|0\rangle = c\,(|0\rangle - c|1\rangle) = c|c\rangle\,. \tag{1.249}$$

It is important to stress that, contrary to the bosonic coherent state, the fermionic coherent state cannot be normalized to 1. In particular one finds

$$\langle c|c\rangle = e^{\bar{c}c} = 1 + \bar{c}c \tag{1.250}$$

because

$$\langle c|c\rangle = (\langle 0| - \bar{c}\langle 1|)\,(|0\rangle - c|1\rangle) = \langle 0|0\rangle + \bar{c}c\langle 1|1\rangle \tag{1.251}$$

having $\langle 0|0\rangle = \langle 1|1\rangle = 1$ and $\langle 1|0\rangle = \langle 0|1\rangle = 0$. Given two fermionic coherent states $|c\rangle = |0\rangle - c|1\rangle$ and $|c'\rangle = |0\rangle - c'|1\rangle$, it is then straightforward to find

$$\langle c|c'\rangle = e^{\bar{c}\,c'} = 1 + \bar{c}\,c' \; . \tag{1.252}$$

Moreover, the spectral resolution of the identity in terms of fermionic coherent states reads

$$\hat{1} = \int d\bar{c}\; dc\, e^{-\bar{c}c}\, |c\rangle\langle c| \; . \tag{1.253}$$

In fact, we have

$$\begin{aligned} \int d\bar{c}\; dc\, e^{-\bar{c}c}\, |c\rangle\langle c| &= \int d\bar{c}\; dc\,(1 - \bar{c}c)\,(|0\rangle - \bar{c}|1\rangle)(\langle 0| - c\langle 1|) \\ &= |0\rangle\langle 0| + |1\rangle\langle 1| - \hat{1} \; . \end{aligned} \tag{1.254}$$

The algebra of Grassmann numbers can be extended in the case of multi-mode fermionic ladder operators $\hat{c}_\alpha$ and $\hat{c}_\alpha^+$, which satisfy the relations

$$\hat{c}_\alpha \hat{c}_\beta^+ + \hat{c}_\beta^+ \hat{c}_\alpha = \delta_{\alpha,\beta} \; , \qquad \hat{c}_\alpha^2 = (\hat{c}_\alpha^+)^2 = 0 \; . \tag{1.255}$$

In this case the Grassmann numbers c_α and $\bar{c}_\alpha$, such that

$$\hat{c}_\alpha|c_\alpha\rangle = c_\alpha|c_\alpha\rangle \; , \tag{1.256}$$

$$\langle c_\alpha|\hat{c}_\alpha^+ = \bar{c}_\alpha\langle c_\alpha| \; , \tag{1.257}$$

have the following properties

$$c_\alpha \bar{c}_\beta = -\bar{c}_\beta c_\alpha \qquad c_\alpha^2 = \bar{c}_\alpha^2 = 0 \; . \tag{1.258}$$

As we did for the bosonic field, we can introduce the pseudo-classical Grassmann analog of the fermionic field operator by using fermionic coherent states. In this case the Eqs. (1.222)–(1.225) are the same, but $|c_\alpha\rangle$ is the coherent state of the fermionic operator $\hat{c}_\alpha$, and c_α is its Grassmann eigenvalue. Explicitly, the Grassmann analog of the fermionic quantum field operator

$$\hat{\psi}(\mathbf{r}) = \sum_\alpha \phi_\alpha(\mathbf{r})\ \hat{c}_\alpha \tag{1.259}$$

is the Grassmann field

$$\psi(\mathbf{r}) = \sum_\alpha \phi_\alpha(\mathbf{r})\ c_\alpha \tag{1.260}$$

such that

$$\hat{\psi}(\mathbf{r})|\psi\rangle = \psi(\mathbf{r})|\psi\rangle\ , \tag{1.261}$$

where

$$|\psi\rangle = \prod_\alpha |c_\alpha\rangle \tag{1.262}$$

is the fermionic coherent state of the system, $|c_\alpha\rangle$ is the coherent state of the fermionic operator $\hat{c}_\alpha$, and c_α is its Grassmann eigenvalue. Similarly, we also have

$$\hat{\psi}^+(\mathbf{r}) = \sum_\alpha \phi^*_\alpha(\mathbf{r})\ \hat{c}^+_\alpha \tag{1.263}$$

with the corresponding complex conjugate Grassmann field

$$\hat{\bar{\psi}}(\mathbf{r}) = \sum_\alpha \phi^*_\alpha(\mathbf{r})\ \bar{c}_\alpha\ . \tag{1.264}$$

1.11 Field Theory at Finite Temperature

In this section we discuss elements of quantum statistical mechanics at thermal equilibrium following the approach introduced by Josiah Willard Gibbs in 1902. This approach is based on the so-called statistical ensembles. Because quantum statistical mechanics is more general than classical statistical mechanics, and it reproduces all the results of thermodynamics in the high-temperature regime, we will consider here directly the quantum statistical mechanics. There are generally many microscopic quantum states which correspond to the same macroscopic thermodynamical state characterized, for instance by the triad (P, V, T), where P is the pressure, V is the volume, and T is the temperature. Gibbs imagined a set of systems in different microscopic quantum states, but all in the same macroscopic state, and he indicated this set with the name statistical ensemble. According to Gibbs, at least three types of statistical ensembles can be considered, corresponding to three types of macroscopic thermodynamic systems:

a) **Microcanonical**: mechanically isolated system without heat exchanges. In this case the internal energy E, the volume V, and the number of particles N are fixed

b) **Canonical**: mechanically isolated system but with heat exchanges. In this case the temperature T, the volume V, and the number N of particles are fixed. The internal energy E is not conserved, and the energy can flow between the thermodynamic system under consideration and the external environment (thermal bath) that surrounds it.

c) **Grand Canonical**: system that can exchange work, heat and particles with the outside. In this case the temperature T, the volume V, and the chemical potential μ are fixed. In this case, the number N of particles and the internal energy E are not preserved.

1.11.1 Quantum Statistical Ensembles

We consider a many-body quantum system characterized by the Hamiltonian $\hat{H}$. We do not specify the form of this Hamiltonian operator because this is not necessary in order to develop a completely general treatment.

1.11.1.1 Quantum Microcanonical Ensemble

In the microcanonical ensemble, the quantum many-body system in a volume V has a fixed number N of particles and also a fixed energy E. The key quantity is the density of states given by

$$W = W_0 \, Tr[\delta \left(E - \hat{H}\right)] , \tag{1.265}$$

where Tr means the trace, i.e., the sum of all diagonal elements, $\delta(x)$ is the Dirac delta function and W_0 is an arbitrary constant with the units of an energy such that W is dimensional. The thermal average of an observable described by the self-adjoint operator $\hat{A}$ is defined as

$$\langle \hat{A} \rangle = \frac{W_0}{W} Tr[\hat{A} \, \delta \left(E - \hat{H}\right)] , \tag{1.266}$$

The connection between statistical mechanics and thermodynamics is given by the formula

$$S = k_B \ \ln(W) , \tag{1.267}$$

which provides the entropy S as a function of energy E, volume V, and number N of particles. Eq. (1.267) is exactly the Boltzmann's entropy formula. Ludwig Boltzmann first developed the formula between 1872 and 1875, and Max Planck changed it into its present form around 1900.

1.11.1.2 Quantum Canonical Ensemble

In the canonical ensemble the quantum system in a volume V has a fixed number N of particles and a fixed temperature T. The key quantity is the canonical partition function $\mathcal{Z}_N$ given by

$$\mathcal{Z}_N = Tr[e^{-\beta \hat{H}}] \tag{1.268}$$

with $\beta = 1/(k_B T)$. The thermal average of an observable $\hat{A}$ is defined as

$$\langle \hat{A} \rangle = \frac{1}{\mathcal{Z}_N} Tr[\hat{A} \ e^{-\beta \hat{H}}] . \tag{1.269}$$

The connection with thermodynamics is given by the formula

$$\mathcal{Z}_N = e^{-\beta F} , \tag{1.270}$$

which provides the Helmholtz free energy F as a function of temperature T, volume V, and number N of particles.

1.11.1.3 Quantum Grand Canonical Ensemble

In the grand canonical ensemble the quantum system in a volume V has a fixed temperature T and a fixed chemical potential μ. The key quantity is the grand canonical partition function given by

$$\mathcal{Z} = Tr\big[e^{-\beta(\hat{H}-\mu\hat{N})}\big] , \tag{1.271}$$

with μ the chemical potential and $\hat{N}$ the number operator. The thermal average of an observable $\hat{A}$ is defined as

$$\langle \hat{A} \rangle = \frac{1}{\mathcal{Z}} Tr\big[\hat{A}\, e^{-\beta(\hat{H}-\mu\hat{N})}\big] . \tag{1.272}$$

The connection with equilibrium thermodynamics is given by

$$\mathcal{Z} = e^{-\beta\Omega} , \tag{1.273}$$

which introduces the grand potential Ω as a function of temperature T, volume V, and chemical potential μ.

It is important to underline that the grand canonical ensemble is the most suitable for quantum field theory, both because in the relativistic context the number of particles is not conserved and also because it is easier to make calculations using this statistical ensemble. Let us analyze in more detail the connection between statistical mechanics and thermodynamics in this ensemble. As previously discussed, thermodynamics dictates that

$$\Omega = E - T\,S - \mu\,N , \tag{1.274}$$

where the internal energy E is directly related to the thermal average of the Hamiltonian $\hat{H}$ by the formula

$$E = \langle \hat{H} \rangle . \tag{1.275}$$

The entropy S satisfies the relation

$$S = -k_B \big\langle \ln \Big(\frac{e^{-\beta(\hat{H}-\mu\hat{N})}}{\mathcal{Z}} \Big) \big\rangle . \tag{1.276}$$

Finally, the thermodynamical total number N of particles is given by

$$N = \langle \hat{N} \rangle . \tag{1.277}$$

1.11.2 Bosons and Fermions at Finite Temperature

Let us consider the non-interacting matter field in thermal equilibrium with a bath at the temperature T. The relevant quantity to calculate all thermodynamical properties of the system is the grand-canonical partition function $\mathcal{Z}$, given by

$$\mathcal{Z} = Tr[e^{-\beta(\hat{H}-\mu\hat{N})}] \tag{1.278}$$

where $\beta = 1/(k_B T)$ with k_B the Boltzmann constant,

$$\hat{H} = \sum_{\alpha} \epsilon_\alpha \, \hat{N}_\alpha \, , \tag{1.279}$$

is the quantum Hamiltonian,

$$\hat{N} = \sum_{\alpha} \hat{N}_\alpha \tag{1.280}$$

is the total number operator, and μ is the chemical potential, fixed by the conservation of the average particle number. This implies that

$$\begin{aligned}
\mathcal{Z} &= \sum_{\{n_\alpha\}} \langle \dots n_\alpha \dots | e^{-\beta(\hat{H}-\mu\hat{N})} | \dots n_\alpha \dots \rangle \\
&= \sum_{\{n_\alpha\}} \langle \dots n_\alpha \dots | e^{-\beta\sum_\alpha(\epsilon_\alpha-\mu)\hat{N}_\alpha} | \dots n_\alpha \dots \rangle \\
&= \sum_{\{n_\alpha\}} e^{-\beta\sum_\alpha(\epsilon_\alpha-\mu)n_\alpha} = \sum_{\{n_\alpha\}} \prod_\alpha e^{-\beta(\epsilon_\alpha-\mu)n_\alpha} \\
&= \prod_\alpha \sum_{n_\alpha} e^{-\beta(\epsilon_\alpha-\mu)n_\alpha} \\
&= \prod_\alpha \sum_{n=0}^{\infty} e^{-\beta(\epsilon_\alpha-\mu)\,n} = \prod_\alpha \frac{1}{1-e^{-\beta(\epsilon_\alpha-\mu)}} \quad \text{for bosons} \quad (1.281) \\
&= \prod_\alpha \sum_{n=0}^{1} e^{-\beta(\epsilon_\alpha-\mu)\,n} = \prod_\alpha \left(1+e^{-\beta(\epsilon_\alpha-\mu)}\right) \quad \text{for fermions}
\end{aligned}$$

(1.282)

Quantum statistical mechanics dictates that the thermal average of any operator $\hat{A}$ is obtained as

$$\langle \hat{A} \rangle = \frac{1}{\mathcal{Z}} Tr[\hat{A}\, e^{-\beta(\hat{H}-\mu\hat{N})}] . \quad (1.283)$$

Let us suppose that $\hat{A} = \hat{H}' = \hat{H} - \mu\hat{N}$, it is then quite easy to show that

$$\begin{aligned} \langle \hat{H}' \rangle &= \frac{1}{\mathcal{Z}} Tr[(\hat{H} - \mu\hat{N})\, e^{-\beta(\hat{H}-\mu\hat{N})}] \\ &= -\frac{\partial}{\partial\beta} \ln\left(Tr[e^{-\beta(\hat{H}-\mu\hat{N}}]\right) = -\frac{\partial}{\partial\beta} \ln(\mathcal{Z}) . \quad (1.284) \end{aligned}$$

By using Eq. (1.281) or Eq. (1.282), we immediately obtain

$$\ln(\mathcal{Z}) = \mp \sum_{\alpha} \ln\left(1 \mp e^{-\beta(\epsilon_\alpha - \mu)}\right) , \quad (1.285)$$

where $-$ is for bosons and $+$ is for fermions, and finally from Eq. (1.284), we get

$$\langle \hat{H} \rangle = \sum_{\alpha} \epsilon_\alpha \langle \hat{N}_\alpha \rangle , \quad (1.286)$$

with

$$\langle \hat{N} \rangle = \sum_{\alpha} \frac{1}{e^{\beta(\epsilon_\alpha - \mu)} \mp 1} . \quad (1.287)$$

Notice that the zero-temperature limit, i.e., $\beta \to \infty$, for fermions we have

$$\langle \hat{N} \rangle = \sum_{\alpha} \Theta\left(\mu - \epsilon_\alpha\right) , \quad (1.288)$$

where $\Theta(x)$ is the Heaviside step function, and the chemical potential μ at zero temperature is nothing but the Fermi energy ϵ_F, i.e., $\epsilon_F = \mu(T = 0)$. Instead, in the high-temperature regime, where $\beta(\epsilon_\alpha - \mu) \gg 1$, Eq. (1.287) becomes

$$\langle \hat{N} \rangle = \sum_{\alpha} e^{-\beta(\epsilon_\alpha - \mu)} , \quad (1.289)$$

that is exactly the Maxwell-Boltzmann distribution. Thus, when the temperature T is such that the highly excited energy levels ϵ_α are occupied, then both Bose-Einstein and Fermi-Dirac distributions reduce to the Maxwell-Boltzmann one.

1.12 Solved Problems

Problem 1.1
Calculate the equation of motion of the Lagrangian

$$L = \frac{m}{2}\dot{q}^2 - \frac{g}{4}\,q^4$$

where $q(t)$ is the dynamical variable, m is the mass, and g is a parameter of the quartic potential energy $U(q) = gq^4/4$.

Solution
We immediately calculate the following expressions

$$\begin{aligned} \frac{\partial L}{\partial q} &= -g\,q^3 \,, \\ \frac{\partial L}{\partial \dot{q}} &= m\dot{q} \,, \\ \frac{d}{dt}\frac{\partial L}{\partial \dot{q}} &= m\ddot{q} \,. \end{aligned}$$

Then, taking into account the Euler-Lagrange equation

$$\frac{d}{dt}\frac{\partial L}{\partial \dot{q}} - \frac{\partial L}{\partial q} = 0 \,,$$

for our specific Lagrangian we get

$$m\ddot{q} + g\,q^3 = 0 \,.$$

Problem 1.2
Given a generic operator $\hat{A}$ and its adjoint $\hat{A}^+$, show that the eigenvalues λ of the operator $\hat{A}^+\hat{A}$ are non-negative.

Solution
The eigenvalue equation of $\hat{A}^+\hat{A}$ reads

$$\hat{A}^+\hat{A}|\lambda\rangle = \lambda|\lambda\rangle \,.$$

We can then write

$$\langle\lambda|\hat{A}^+\hat{A}|\lambda\rangle = \lambda\langle\lambda|\lambda\rangle = \lambda ,$$

because we choose the eigenstate $|\lambda\rangle$ normalized to one. We also have

$$\langle\lambda|\hat{A}^+\hat{A}|\lambda\rangle = (\hat{A}|\lambda\rangle)^+(\hat{A}|\lambda\rangle) = |(\hat{A}|\lambda\rangle)|^2$$

Consequently, we obtain

$$\lambda = |(\hat{A}|\lambda\rangle|^2 \geq 0 .$$

Problem 1.3

Consider the quantum Hamiltonian of the one-dimensional harmonic oscillator

$$\hat{H} = \frac{\hat{p}^2}{2m} + \frac{1}{2}m\omega^2\hat{x}^2 .$$

By using the properties of the annihilation operator

$$\hat{a} = \sqrt{\frac{m\omega}{2\hbar}}\left(\hat{x} + \frac{i}{m\omega}\hat{p}\right) ,$$

calculate the ground state wavefunction.

Solution

First of all we remind that $\hat{p} = -i\hbar d/dx$. Moreover, we have

$$\langle x|\hat{a} = \sqrt{\frac{m\omega}{2\hbar}}\left(x + \frac{\hbar}{m\omega}\frac{d}{dx}\right)\langle x| .$$

In addition, from the basic property

$$\hat{a}|0\rangle = 0 ,$$

we get

$$\langle x|\hat{a}|0\rangle = \sqrt{\frac{m\omega}{2\hbar}}\left(x + \frac{\hbar}{m\omega}\frac{d}{dx}\right)\langle x|0\rangle = 0 ,$$

where $\psi_0(x) = \langle x|0\rangle$ is the ground state wavefunction. Thus,

$$\sqrt{\frac{m\omega}{2\hbar}}\left(x + \frac{\hbar}{m\omega}\frac{d}{dx}\right)\psi_0(x) = 0 .$$

By using the separation of variables, we deduce

$$\psi_0(x) = \frac{1}{\pi^{1/4}\ell^{1/2}}\, e^{-\frac{x^2}{2\ell^2}} .$$

having introduced the characteristic length

$$\ell = \sqrt{\frac{\hbar}{m\omega}}$$

and imposing the normalization

$$\int dx\ |\psi_0(x)|^2 = 1 .$$

Problem 1.4
Discuss examples where the distribution of photons in a single mode is Planckian, or Maxwellian, or Poissonian.

Solution
Let us consider a single mode of the electromagnetic field characterized by wavevector $\mathbf{k}$, polarization s, and angular frequency $\omega_k = ck$ with $k = |\mathbf{k}|$. At thermal equilibrium with a bath of temperature T, the distribution of photons is Planckian

$$\langle \hat{N}_{\mathbf{k}s} \rangle = \frac{1}{e^{\hbar\omega_k/(k_B T)} - 1} ,$$

that is the Bose-Einstein distribution with a null chemical potential μ. This distribution becomes Maxwellian, i.e., a Maxwell-Boltzmann,

$$\langle \hat{N}_{\mathbf{k}s} \rangle = e^{-\hbar\omega_k/(k_B T)}$$

for $\hbar\omega_k \gg k_B T$. Instead, if the photons are in the coherent state $|a_{\mathbf{r}k}\rangle$, that is the eigenstate of the annihilation operator $\hat{a}_{\mathbf{k}s}$, we have

$$\langle a_{\mathbf{k}s} | \hat{N}_{\mathbf{b}s} | a_{\mathbf{k}s} \rangle = |a_{\mathbf{k}s}|^2 .$$

In this case the probability of finding the coherent state $|a_{\mathbf{k}s}\rangle$ in the Fock state $|n_{\mathbf{k}s}\rangle$ is Poissonian

$$|\langle a_{\mathbf{k}s} | n_{\mathbf{k}s} \rangle|^2 = e^{-|a_{\mathbf{k}s}|^2}\, \frac{|a_{\mathbf{k}s}|^{2n_{\mathbf{k}s}}}{n_{\mathbf{k}s}!} .$$

Problem 1.5
Calculate the grand canonical partition function $\mathcal{Z}$ and the average number $\langle \hat{N} \rangle$ of particles for a system of non-interacting two-spin-component fermions that can occupy only two single-particle energy levels with energies $\epsilon_1 = 0$ and $\epsilon_2 = \Delta$.

Solution
In the grand canonical ensemble the requested partition function Z is given by

$$\begin{aligned}\mathcal{Z} &= (1 + e^{-\beta(\epsilon_1 - \mu)})^2 \, (1 + e^{-\beta(\epsilon_2 - \mu)})^2 \\ &= (1 + e^{-\beta\mu})^2 \, (1 + e^{-\beta(\Delta - \mu)})^2 = e^{-\beta\Omega} \, ,\end{aligned}$$

where the number 2 in the exponents takes into account that each level can hold at most two fermions due to spin degeneracy, and Ω is the grand potential. As usual, $\beta = 1/(k_B T)$ with k_B the Boltzmann constant and T the temperature. Instead, the average number of particles reads

$$\langle \hat{N} \rangle = - \left(\frac{\partial \Omega}{\partial \mu} \right)_T = \frac{2}{e^{-\beta\mu} + 1} + \frac{2}{e^{\beta(\Delta - \mu)}} \, .$$

As $T \to 0$ the Fermi-Dirac distribution becomes a step function. It follows that:

$$\langle \hat{N} \rangle = \begin{cases} 4 & \text{for} \quad \mu > \Delta \\ 2 & \text{for} \quad 0 < \mu < \Delta \\ 0 & \text{for} \quad \mu < 0 \end{cases} \, .$$

Further Reading

Classic books of analytical mechanics are
– H. Goldstein, C.P. Poole, and J. Safko, Classical Mechanics (Pearson, London, 2011)
– V.I. Arnold, Mathematical Methods of Classical Mechanics (Springer, Berlin, 1989).
A deep discussion of second quantization for many-body problems can be found in
– A.L. Fetter and J.D. Walecka, Quantum Theory of Many-Particle Systems (Dover, New York, 2003).
The canonical quantum field theory for non-relativistic bosonic and fermionic particles is explained in many textbooks. Among them
– L. Salasnich, Quantum Physics of Light and Matter: Photons, Atoms, and Strongly Correlated Systems (Springer, Cham, 2017).
– N. Dupuis, Field Theory of Condensed Matter and Ultracold Gases, vol. 1 (World Scientific, Singapore, 2023).

2

Path Integral for Particles

In this chapter we show that it is possible to describe quantum mechanics by using path integrals. This idea was depeloped by Richard Feynman on the basis of an earlier intuition of Paul Dirac. We discuss the path integral approach starting with the problem of a single quantum particle in the presence of an external potential. We introduce the concept of the saddle point and the stationary phase approximation, which is solved by using Fresnel integrals. We show that Gaussian integrals are crucial in the imaginary time formulation of the path integral, which is the appropriate framework for studying quantum systems at thermal equilibrium. Finally, we analyze the quantum tunneling problem, deriving an elegant formula for the tunneling probability based on the simple, but meaningful, saddle point approximation.

2.1 Introduction

Let us introduce this chapter with a very short philosophical digression. Reductionism asserts that the nature of complex structures always can be reduced and explained by simpler structures and fundamentals. Just like determinism, reductionism has also always encountered strong hostility in the philosophical field. As discussed in the previous chapter, in the last years, reductionism has been questioned not only in philosophy but also by authoritative exponents of basic scientific disciplines, such as biology and chemistry. Also eminent physicists, mainly operating in the condensed matter sectors and statistical mechanics, have addressed criticisms of the reductionist vision of reality, usually supported by elementary particle physicists. The theory of complexity is the

DOI: 10.1201/9781003588337-2

modern current of thought that believes, as did Aristotle, that the whole is more than the parts that constitute it. This is a typical statement of a known philosophical theory as holism, often contrasted with reductionism. However, holism goes further, stating that the whole determines how the individual parts behave. Complexity theory does not go that far, but notes that some phenomena (called collective or emergent phenomena) are the result of the interaction of many parts strongly related.

The science of complexity has its roots in cybernetics and self-organization theory. Cybernetics studies the phenomena of self-regulation and communication, both in natural organisms and in artificial systems. It is therefore a field of interdisciplinary studies, between science and engineering. The term cybernetics was coined in 1947 by the mathematician Norbert Wiener. Wiener played a crucial role also in the development of the path integral discussed in this section. In fact, Norbert Wiener, who developed the Wiener integral to address diffusion and Brownian motion problems in the early 1920s, is acknowledged for having developed the fundamental concept of the path integral formulation. Paul Dirac expanded on this concept by using the Lagrangian in quantum mechanics in 1933. Richard Feynman created the entire technique in 1948. The original goal was to use a Lagrangian (instead of a Hamiltonian) as the basis for the entire quantum mechanics.

2.2 Propagator and Feynman Path Integral

To warm up, we consider the quantum mechanics of one particle moving on a line. The general form of the Hamiltonian for this single particle is expressed as

$$\hat{H} = \frac{\hat{p}^2}{2m} + U(\hat{x}), \tag{2.1}$$

where m is the mass of the particle, $\hat{p}$ is the momentum operator, and $U(\hat{x})$ is the external potential energy operator. As discussed in the previous chapter, the time evolution of a quantum state from the initial time t_I to the final time t_F is driven by the unitary operator

$$\hat{U}(t_F - t_I) = e^{-\frac{i}{\hbar}\hat{H}(t_F - t_I)} \; . \tag{2.2}$$

The matrix element of this evolution operator

$$K(x_F, t_F|x_I, t_I) = \langle x_F|\hat{U}(t_F - t_I)|x_I\rangle = \langle x_F t_F|x_I t_I\rangle \qquad (2.3)$$

is usually called the *propagator* or, alternatively, the *kernel*. This propagator gives the probability amplitude of finding the particle at the final position x_F at final time t_F conditioned by having the particle at the initial position x_I at the initial time t_I. Notice that

$$K(x_F, t|x_I, t) = \langle x_F t|x_I t\rangle = \langle x_F|x_I\rangle = \delta(x_F - x_I) \qquad (2.4)$$

with $\delta(x)$ the Dirac delta function. Clearly

$$P_T(x_F, t_F|x_I, t_I) = |K(x_F, t|x_I, t)|^2 \qquad (2.5)$$

is the quantum transition probability (conditional probability) of finding a particle in the position x_F at the final time t_F with the condition (constraint) of being in the position x_I at the initial time t_I.

We observe that, knowing the propagator $K(x, t|x', t')$, the wavefunction $\psi(x, t)$ of the particle can be determined as follows

$$\psi(x, t) = \int_{-\infty}^{+\infty} dx'\, K(x, t|x', 0)\, \psi_0(x')\ , \qquad (2.6)$$

where $\psi_0(x) = \psi(x, 0)$ is the initial wavefunction, i.e., the wavefunction at $t = 0$. In fact,

$$\begin{aligned}
\psi(x, t) &= \langle xt|\psi\rangle = \langle xt|\hat{1}|\psi\rangle = \langle xt| \int_{-\infty}^{+\infty} dx' |x'0\rangle\langle x'0|\psi\rangle \\
&= \int_{-\infty}^{+\infty} dx'\, \langle xt|x'0\rangle\, \langle x'0|\psi\rangle \\
&= \int_{-\infty}^{+\infty} dx' \langle xt|x'0\rangle\, \psi_0(x') \\
&= \int_{-\infty}^{+\infty} dx'\, K(x, t|x', 0)\, \psi_0(x') \qquad (2.7)
\end{aligned}$$

with $K(x, t|x', 0) = \langle xt|x'0\rangle$. Thus, the time-dependent Schrödinger equation

$$i\hbar \frac{\partial}{\partial t}\psi(x, t) = \left(-\frac{\hbar^2}{2m}\frac{\partial^2}{\partial x^2} + U(x)\right)\psi(x, t) \qquad (2.8)$$

with the initial condition $\psi_0(x)$ is basically equivalent to Eq. (2.6). In functional analysis, Eq. (2.6) is called *weak form*, while Eq. (2.8) is the *strong form* of the problem involving partial derivatives. Eq. (2.6) justifies the word "kernel" for the propagator. We observe that also the quantum propagator $K(x,t|x',t')$ satisfies the time-dependent Schrödinger equation, in particular

$$\left(i\hbar\frac{\partial}{\partial t} + \frac{\hbar^2}{2m}\frac{\partial^2}{\partial x^2} - U(x) \right) K(x,t|x',t') = 0 \; . \tag{2.9}$$

It is quite simple to prove this formula by using the definition of the quantum propagator in terms of the unitary evolution operator. In particular,

$$\begin{aligned} i\hbar\frac{\partial}{\partial t}K(x,t|x',t') &= i\hbar\frac{\partial}{\partial t}\langle x|e^{-\frac{i}{\hbar}\hat{H}(t-t')}|x'\rangle = i\hbar\langle x|\frac{\partial}{\partial t}e^{-\frac{i}{\hbar}\hat{H}(t-t')}|x'\rangle \\ &= \langle x|\hat{H}e^{-\frac{i}{\hbar}\hat{H}(t-t')}|x'\rangle \\ &= \left(-\frac{\hbar^2}{2m}\frac{\partial^2}{\partial x^2} + U(x) \right) \langle x|e^{-\frac{i}{\hbar}\hat{H}(t-t')}|x'\rangle \\ &= \left(-\frac{\hbar^2}{2m}\frac{\partial^2}{\partial x^2} + U(x) \right) K(x,t|x',t') \; . \end{aligned} \tag{2.10}$$

Another relevant result is the following decomposition of the propagator

$$K(x,t|x',t') = \sum_n \phi_n(x)\, \phi_n^*(x')\, e^{-\frac{i}{\hbar}E_n(t-t')} \; , \tag{2.11}$$

where

$$\hat{H}|\phi_n\rangle = E_n|\phi_n\rangle \tag{2.12}$$

with $\phi_n(x) = \langle x|\phi_n\rangle$. Eq. (2.11) can be proven taking into acount the spectral decomposition of the identity operator

$$\hat{1} = \sum_n |\phi_n\rangle\langle\phi_n| \tag{2.13}$$

as follows

$$\begin{aligned} K(x,t|x',t') &= \langle xt|x't'\rangle = \langle xt|\hat{1}|x't'\rangle = \langle xt| \left(\sum_n |\phi_n\rangle\langle\phi_n| \right) |x't'\rangle \\ &= \sum_n \langle xt|\phi_n\rangle\, \langle\phi_n|x't'\rangle \\ &= \sum_n e^{-\frac{i}{\hbar}E_n t}\langle x|\phi_n\rangle\, e^{\frac{i}{\hbar}E_n t'}\, \langle x'|\phi_n\rangle \end{aligned} \tag{2.14}$$

observing that

$$\langle xt|\phi_n\rangle = \langle x|e^{-\frac{i}{\hbar}\hat{H}t}|\phi_n\rangle = e^{-\frac{i}{\hbar}E_n t}\langle x|\phi_n\rangle = e^{-\frac{i}{\hbar}E_n t}\,\phi_n(x) \tag{2.15}$$

but also

$$|xt\rangle = e^{\frac{i}{\hbar}\hat{H}t}|x\rangle \; . \tag{2.16}$$

The physical interpretation of the quantum propagator as a functional integral over all the possible paths weighted by the exponential of the classical action was suggested by Feynman in 1948. The amazing path-integral result obtained by Feynman is encoded in the formula

$$K(x_F, t_F|x_I, t_I) = \int_{x(t_I)=x_I}^{x(t_F)=x_F} \mathcal{D}[x]\, e^{\frac{i}{\hbar}S[x]} \; , \tag{2.17}$$

where

$$S[x] = \int_{t_I}^{t_F} dt\, \left(\frac{m}{2}\dot{x}(t)^2 - U(x(t))\right) \tag{2.18}$$

is the action functional of the problem. Notice that a very simplified version of Eq. (2.17) appears in a paper of Dirac published in 1933. All the Feynman paths $x(t)$ must satisfy the same boundary conditions: $x(t_I) = x_I$ and $x(t_F) = x_F$. The classical path $x_c(t)$, i.e., the path $x(t)$ that extremizes the action, is continuous and differentiable, while typical paths $x(t)$ of the Feynman path integral are not necessarily differentiable. Mathematically speaking, differentiable paths have zero measure in the set of Feynman paths.

2.2.1 Gaussian vs Fresnel Integrals

The construction and the various applications of the Feynman path integral are based on the Fresnel and Gaussian integrals. A typical Gaussian integral is

$$\int_{-\infty}^{+\infty} dx\, e^{-\frac{a}{2}x^2+bx} = \sqrt{\frac{2\pi}{a}}\; e^{\frac{b^2}{2a}} \; , \tag{2.19}$$

where x is the real variable of integration and a is a non-negative real number. The proof of this result can be obtained by a change of a variable after proving the most relevant Gaussian formula

$$\int_{-\infty}^{+\infty} dx\, e^{-x^2} = \sqrt{\pi}\,. \tag{2.20}$$

The proof goes as follows

$$\begin{aligned}
\int_{-\infty}^{+\infty} dx\, e^{-x^2} &= \left(\int_{-\infty}^{+\infty} dx\, e^{-x^2} \int_{-\infty}^{+\infty} dy\, e^{-y^2}\right)^{1/2} \\
&= \left(\int_{-\infty}^{+\infty}\int_{-\infty}^{+\infty} dx\, dy\, e^{-(x^2+y^2)}\right)^{1/2} \\
&= \left(\int_{0}^{+\infty} dr\, r \int_{0}^{2\pi} d\phi\, e^{-r^2}\right)^{1/2} \\
&= \left(2\pi \int_{0}^{+\infty} dr\, r\, e^{-r^2}\right)^{1/2} \\
&= \left(2\pi \left[-\frac{1}{2}e^{-r^2}\right]_{0}^{+\infty}\right)^{1/2} = \sqrt{\pi}
\end{aligned} \tag{2.21}$$

introducting the polar coordinates $r = \sqrt{x^2+y^2}$ and $\phi = \arctan(y/x)$.

Another typical Gaussian integral is

$$\int_{-\infty}^{+\infty} dx_1\, e^{-\frac{a}{2}[(x_2-x_1)^2+(x_1-x_0)^2]} = \sqrt{\frac{\pi}{a}}\; e^{-\frac{a}{4}(x_2-x_0)^2}\,, \tag{2.22}$$

where a is again a non-negative real number. The integrals (2.19) and (2.22) become Fresnel integrals if a is a complex number with a non-zero imaginary part. In this case, Eqs. (2.19) and (2.22) are still valid in the spirit of the analytic continuation. However, while the integrand of these Gaussian integrals goes to zero as $x \to \pm\infty$, in the case of the Fresnel integrals, the integrand is a complex function, whose real and imaginary parts have an oscillatory behavior, and the convergence is ensured if the real part of a is positive. In fact the Fresnel integrals belong to the family of oscillatory integrals. Not even to say, the mathematical rigour and the numerical convergence of Gaussian integrals is much less problematic than

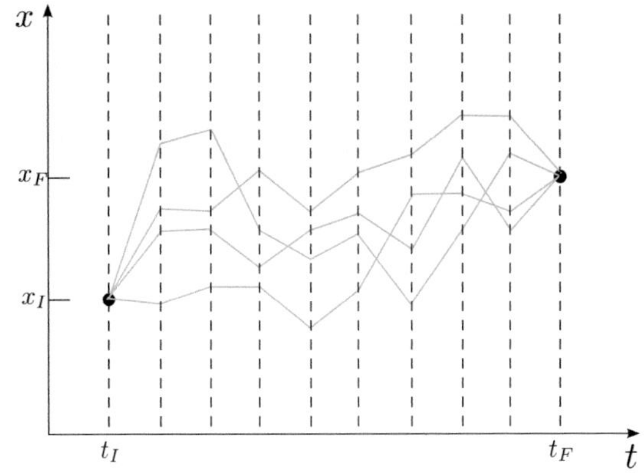

Figure 2.1
Schematic representation of the paths connecting the initial and final states. In the discretization with N temporal intervals, we have fixed boundary conditions: $t_I = t_0$ with $x_I = x_0$ and $t_F = t_N$ with $x_F = x_N$.

that of Fresnel integrals. As we will see, the Feynman path integral is constructed with Fresnel integrals, which, however, become Gaussian integrals if a Wick rotation is carried out from real time to imaginary time. In the case of multidimensional integrals which involve the positive definite $N \times N$ matrix $\hat{A}$ and the vector $\mathbf{b}$ with N elements, we have

$$\int_{\mathbb{R}^n} d^N\mathbf{x}\, e^{-\frac{1}{2}\mathbf{x}\cdot\hat{A}\mathbf{x}+\mathbf{b}\cdot\mathbf{x}} = \sqrt{\frac{(2\pi)^N}{\det[\hat{A}]}}\; e^{\frac{1}{2}\mathbf{b}\cdot\hat{A}^{-1}\mathbf{b}}\,, \tag{2.23}$$

where $\det[\hat{A}]$ is the determinant of the matrix $\hat{A}$.

2.2.2 Short-Time Propagator

We now want to prove Eq. (2.17). First, we subdivide the interval $[t_I, t_F]$ in N subintervals of period $\Delta t = (t_F - t_I)/N$ and look at the position x_j of the particle at time $t_j = t_I + j\Delta t$, with $j = 0, 1, 2, ..., N$. It follows that $t_0 = t_I$ and $t_N = t_F$ and consequently $x_0 = x_I$ and $x_N = x_F$. See Fig. 2.1. Without integrating over

$x_0 = x_I$ and $x_N = x_F$, we insert $N-1$ identity operators

$$\hat{1} = \int_{-\infty}^{+\infty} dx_j |x_j t_j\rangle\langle x_j t_j| \tag{2.24}$$

in the propagator finding

$$\begin{aligned}
\langle x_F t_F | x_I t_I \rangle &= \langle x_F t_F | \left(\prod_{j=N-1}^{1} \int_{-\infty}^{+\infty} dx_j |x_j t_j\rangle\langle x_j t_j| \right) |x_I t_I\rangle \\
&= \int_{\mathbb{R}^{N-1}} \left(\prod_{n=1}^{N-1} \mathrm{d}x_n \right) \prod_{j=N-1}^{0} \langle x_{j+1} t_{j+1} | x_j t_j \rangle \\
&= \int_{\mathbb{R}^{N-1}} \left(\prod_{n=1}^{N-1} \mathrm{d}x_n \right) \prod_{j=N-1}^{0} \langle x_{j+1} | e^{-\frac{i}{\hbar}\hat{H}\Delta t} | x_j \rangle . \qquad (2.25)
\end{aligned}$$

It is now crucial to carefully analyze the short-time propagator

$$\langle x_{j+1} | e^{-\frac{i}{\hbar}\hat{H}\Delta t} | x_j \rangle = \langle x_{j+1} | e^{-\frac{i}{\hbar}(\hat{T}+\hat{U})\Delta t} | x_j \rangle \tag{2.26}$$

because the kinetic energy operator

$$\hat{T} = \frac{\hat{p}^2}{2m} \tag{2.27}$$

does not commute with the potential energy operator

$$\hat{U} = U(\hat{x}) . \tag{2.28}$$

In order to solve this problem, we use the Trotter-Suzuki decomposition

$$e^{-\frac{i}{\hbar}(\hat{T}+\hat{U})\Delta t} = e^{-\frac{i}{\hbar}\hat{T}\Delta t} e^{-\frac{i}{\hbar}\hat{U}\Delta t} + O(\Delta t^2) \tag{2.29}$$

under the assumption of a very small Δt. In this way, we get

$$\begin{aligned}
\langle x_F t_F | x_I t_I \rangle = &\int_{\mathbb{R}^{N-1}} \left(\prod_{n=1}^{N-1} \mathrm{d}x_n \right) \prod_{j=N-1}^{0} \langle x_{j+1} | e^{-\frac{i}{\hbar}\hat{T}\Delta t} | x_j \rangle \, e^{-\frac{i}{\hbar}U(x_j)\Delta t} \\
&+ O(\Delta t^2) . \qquad (2.30)
\end{aligned}$$

For the short-time propagator of the kinetic energy operator, we insert the following identity

$$\hat{1} = \int_{-\infty}^{+\infty} |p_j\rangle\langle p_j|\, dp_j \tag{2.31}$$

and obtain

$$\begin{aligned}\langle x_{j+1}|e^{-\frac{i}{\hbar}\frac{\hat{p}^2}{2m}\Delta t}|x_j\rangle &= \langle x_{j+1}|e^{-\frac{i}{\hbar}\frac{\hat{p}^2}{2m}\Delta t}\left(\int_{-\infty}^{+\infty} dp_j\, |p_j\rangle\langle p_j|\right)|x_j\rangle \\ &= \int_{-\infty}^{+\infty} \mathrm{d}p_j\; e^{-\frac{i}{\hbar}\frac{p_j^2}{2m}\Delta t}\,\langle x_{j+1}|p_j\rangle\,\langle p_j|x_j\rangle \\ &= \int_{-\infty}^{+\infty} \frac{\mathrm{d}p_j}{2\pi\hbar} e^{-\frac{i}{\hbar}\frac{p_j^2}{2m}\Delta t}\, e^{-\frac{i}{\hbar}p_j(x_{j+1}-x_j)} \end{aligned}\tag{2.32}$$

remembering that the plane wave can be written within the Dirac notation as $\langle x|p\rangle$, i.e.

$$\langle x|p\rangle = \frac{e^{\frac{i}{\hbar}px}}{\sqrt{2\pi\hbar}}\,. \tag{2.33}$$

The Fresnel integral (2.32) can be calculated analytically giving

$$\langle x_{j+1}|e^{-\frac{i}{\hbar}\frac{\hat{p}^2}{2m}\Delta t}|x_j\rangle = \sqrt{\frac{m}{2\pi i\hbar\Delta t}}\, e^{\frac{i}{\hbar}\frac{m}{2}\frac{(x_{j+1}-x_j)^2}{\Delta t}} \tag{2.34}$$

and consequently

$$\begin{aligned}K(x_{j+1}, t_j + \Delta t|x_j, t_j) &= \langle x_{j+1}|e^{-\frac{i}{\hbar}\hat{H}\Delta t}|x_j\rangle \\ &= \sqrt{\frac{m}{2\pi i\hbar\Delta t}}\, e^{\frac{i}{\hbar}\Delta t[\frac{m}{2}\frac{(x_{j+1}-x_j)^2}{\Delta t^2} - U(x_j)]}\end{aligned} \tag{2.35}$$

is the analytical expression of the short-time propagator for $\Delta t \to 0$. This is the key formula connecting the Hamiltonian formulation to the Lagrangian one.

Therefore, at the end we obtain

$$\begin{aligned}K(x_F, t_F|x_I, t_I) = \int_{\mathbb{R}^{N-1}} \Big\{ \left(\prod_{n=1}^{N-1} dx_n\right)\left(\frac{m}{2\pi i\hbar\Delta t}\right)^{N/2} \\ \cdot\, e^{\frac{i}{\hbar}\sum_{j=0}^{N-1}[\frac{m}{2}\frac{(x_{j+1}-x_j)^2}{\Delta t^2} - U(x_j)]\Delta t}\Big\} + O(\Delta t^2)\,.\end{aligned} \tag{2.36}$$

The limit of $\Delta t \to 0$ with $N \to \infty$ leads us to the formal expression (2.17) with (2.18), where

$$\lim_{\substack{\Delta t \to 0 \\ N \to \infty}} \int_{\mathbb{R}^{N-1}} \left(\prod_{n=1}^{N-1} dx_n \right) \left(\frac{m}{2\pi i \hbar \Delta t} \right)^{N/2} \to \int_{x(t_I)=x_I}^{x(t_F)=x_F} \mathcal{D}[x(t)] \tag{2.37}$$

is the specific definition of the path integral (or functional integral) introduced by Feynman. The symbol $x(t)$ denotes the Feynman path $(x_0, x_1, ..., x_{N-1}, x_N)$ in the limit $N \to \infty$. Strictly speaking, this notation does not imply continuity or differentiability of the Feynman path. However, in several applications, one safely works assuming that $x(t)$ is continuous and also differentiable.

For a generic external potential $U(x)$, it is not possible to calculate analytically the Feynman path integral. However, there are cases where analytical results can be obtained. For instance, for the free particle and the harmonic oscillator, the path integral can be computed directly from the sum over discrete trajectories. Alternatively, the same results can be obtained by using the continuous path formulation.

2.2.3 Path Integral in Phase Space

For the sake of completeness, we observe that one could avoid to calculating explicitly the integral of Eq. (2.32). Inserting directly Eq. (2.32) into Eq. (2.25), we get

$$\begin{aligned} \langle x_F t_F | x_I t_I \rangle &= \int_{\mathbb{R}^{2N-1}} \left(\prod_{n=1}^{N-1} \mathrm{d}x_n \right) \left(\prod_{k=0}^{N-1} \frac{\mathrm{d}p_n}{(2\pi\hbar)} \right) \\ &\quad \times e^{\sum_{j=0}^{N-1} \frac{i}{\hbar} \left(p_j (\frac{x_{j+1}-x_j}{\Delta t}) - H(p_j, x_j) \right) \Delta t} , \end{aligned} \tag{2.38}$$

where $H(x_j, p_j) = p_j^2/(2m) + U(x_j)$. Notice that we are integrating over all the p_j that appear in the exponent of the complex exponential. In the continuum limit, remembering that $K(x_F, t_F | x_I, t_I) = \langle x_F t_F | x_I t_I \rangle$, the previous expression formally becomes

$$K(x_F, t_F | x_I, t_I) = \int_{x(t_I)=x_I}^{x(t_F)=x_F} \mathcal{D}[x] \int \mathcal{D}[p] \, e^{\frac{i}{\hbar} S[x,p]} , \tag{2.39}$$

where

$$S[x,p] = \int_{t_I}^{t_F} dt \; [p(t)\dot{x}(t) - H(x(t),p(t))] \qquad (2.40)$$

is the phase-space action functional, with $H(x,p) = p^2/(2m) + U(x)$ in our specific case. Eq. (2.39) is more general with respect to Eq. (2.17) because it could be applied to Hamiltonians which do not have a kinetic term that is quadratic with respect to the linear momentum. However, in high-energy physics, the generalization of Eq. (2.17) is usually adopted to quantize the fields without referring to a Hamiltonian formulation.

2.2.4 From Feynman to Schrödinger

It is possible to derive the Schödinger equation starting from the Feynman's path integral representation of the propagator. First of all we write

$$\psi(x,t+\Delta t) = \int_{-\infty}^{+\infty} dy \, K(x,t+\Delta t|y,t) \, \psi(y,t) \qquad (2.41)$$

remembering the basic properties of the propagator. Here Δt is a very small time interval. Performing the change of variable

$$y = x + \xi \qquad (2.42)$$

we have

$$\psi(x,t) + \Delta t \frac{\partial}{\partial t}\psi(x,t) + ... = \int_{-\infty}^{+\infty} d\xi \, K(x,t+\Delta t|x+\xi,t) \, \psi(x+\xi,t) \qquad (2.43)$$

having performed a Taylor expansion of $\psi(x,t+\Delta)$ at the first order with respect to Δt. Formally we also have

$$\psi(x+\xi,t) = \psi(x,t) + \xi \frac{\partial}{\partial x}\psi(x,t) + \frac{1}{2}\xi^2 \frac{\partial^2}{\partial x^2}\psi(x,t) + ... \qquad (2.44)$$

at the second order in a Taylor expansion of $\psi(x+\xi,t)$ with respect to ξ. We also Taylor-expand the short-time propagator, see Eq.

(2.35), in this way

$$
\begin{aligned}
K(x,t+\Delta t|x+\xi,t) &= \sqrt{\frac{m}{2\pi\hbar i\Delta t}}\, e^{\frac{i}{\hbar}[\frac{\xi^2}{2m\Delta t}-\Delta t\, U(x+\xi)]} \\
&= \sqrt{\frac{m}{2\pi\hbar i\Delta t}}\, e^{\frac{i}{\hbar}\frac{\xi^2}{2m\Delta t}} \left(1-\frac{i}{\hbar}\Delta t\, U(x)+...\right) .
\end{aligned}
\tag{2.45}
$$

Putting everything together and integrating with respect to ξ, we get

$$
\psi(x,t)+\Delta t\frac{\partial}{\partial t}\psi(x,t) = \psi(x,t)+\frac{\Delta t}{i\hbar}U(x)\psi(x,t)-\frac{\Delta t}{i\hbar}\frac{\hbar^2}{2m}\frac{\partial^2}{\partial x^2}\psi(x,t) . \tag{2.46}
$$

A simplification and a simple rearrangement show that this formula is nothing else than the time-dependent Schrödinger equation (2.8). Note that higher-order terms in the expansion (2.43) do to contribute after integration with respect to ξ in the limit $\Delta t \to 0$.

2.2.5 Free Particle

In the case of the free particle, where the Trotter-Suzuki decomposition is not needed because $U(x) = 0$, Eq. (2.36) becomes

$$
K_{\text{free}}(x_F,t_F|x_I,t_I) = \int_{\mathbb{R}^{N-1}} \prod_{n=1}^{N-1} dx_n \sqrt{\frac{m}{2\pi i\hbar\Delta t}}\, e^{\frac{i}{\hbar}\frac{m}{2\Delta t}\sum_{j=0}^{N-1}(x_{j+1}-x_j)^2} \tag{2.47}
$$

without the $O(\Delta t^2)$ term. Thus, in this case, it is not necessary to do the limit $\Delta t \to 0$ and $N \to \infty$. One must, however, calculate $N-1$ Fresnel integrals. The calculation is a bit tedious, but the final result is quite simple, namely

$$
K_{\text{free}}(x_F,t_F|x_I,t_I) = \sqrt{\frac{m}{2\pi i\hbar(t_F-t_I)}}\, e^{\frac{i}{\hbar}\frac{m(x_F-x_I)^2}{2(t_F-t_I)}} , \tag{2.48}
$$

because after $N-1$ Fresnel integrations, see Eq. (2.22), Δt becomes $\Delta t + (N-1)\Delta t = N\Delta t = t_F - t_I$. Formally, in the exponential of Eq. (2.48) there is the action functional of the free particle

$$
S[x(t)] = \int_{t_I}^{t_F} dt\, \frac{m}{2}\dot{x}(t)^2 \tag{2.49}
$$

calculated along the classical trajectory

$$x_c(t) = x_I + \frac{x_F - x_I}{t_F - t_I} t \tag{2.50}$$

which minimizes the action $S[x]$, i.e.

$$S[x_c(t)] = \frac{m}{2} \frac{(x_F - x_I)^2}{(t_F - t_I)} \ . \tag{2.51}$$

2.2.6 Free Particle: Simpler Derivation

It is interesting to observe that the propagator of the free particle can be calculated more easily without adopting the path integral formulation. In particular, we have

$$\begin{aligned} K_{\text{free}}(x_F, t_F|x_I, t_I) &= \langle x_F|e^{-\frac{i}{\hbar}\frac{\hat{p}^2}{2m}(t_F-t_I))}|x_I\rangle \\ &= \langle x_F|e^{-\frac{i}{\hbar}\frac{\hat{p}^2}{2m}(t_F-t_I)}|\hat{1}|x_I\rangle \\ &= \langle x_F|e^{-\frac{i}{\hbar}\frac{\hat{p}^2}{2m}(t_F-t_I)}\left(\int_{-\infty}^{+\infty} \mathrm{d}p|p\rangle\langle p|\right)|x_I\rangle \\ &= \int_{-\infty}^{+\infty} \frac{\mathrm{d}p}{2\pi\hbar} e^{-\frac{i}{\hbar}\frac{p^2}{2m}(t_F-t_I)}\, e^{-\frac{i}{\hbar}p(x_F-x_I)} \end{aligned} \tag{2.52}$$

and performing the Fresnel integral, we recover Eq. (2.48).

As previously discussed, having the propagator one immediately gets the time-dependent wavefunction $\psi(x,t)$ given the initial wavefunction $\psi_0(x) = \psi(x,0)$. For the free particle, from Eq. (2.48) we obtain

$$\psi(x,t) = \int_{-\infty}^{+\infty} dx' \sqrt{\frac{m}{2\pi i\hbar t}}\, e^{\frac{i}{\hbar}\frac{m(x-x')^2}{2t}}\, \psi_0(x') \ , \tag{2.53}$$

which is quite an elegant result.

2.3 Saddle Point and Stationary Phase

In full generality the evaluation of the propagator of Eq. (2.17) is a very difficult task. However, the intuition suggests that the main contribution to the overall integral is obtained for a stationary value of the action. In fact, paths that are far from the stationary one are together contributing with weights that are oscillating in the complex unitary circle, so we expect a mutual cancellation of the corresponding values. If we consider instead the paths near the stationary argument of the action, their effects sum up. This stationarity principle is stated by imposing the first variation of S to zero

$$\delta S[x] = 0 \, , \tag{2.54}$$

that is the least action principle. From this point of view, one could say that the least action principle of classical mechanics is embedded in Feynman's path integral approach to quantum mechanics, and it explicitly appears in the limit $\hbar \to 0$. Let $x_c(t)$ be a solution of this equation, such that $x_c(t_I) = x_I$ and $x_c(t_F) = x_F$. This function must satisfy the Euler-Lagrange equation: indeed the trajectory $x_c(t)$ with $t \in [t_I, t_F]$ is often called a *classical solution.* To the leading order, one can express the overall functional integral just keeping the stationary value

$$K(x_F, t_F | x_I, t_I) = e^{\frac{i}{\hbar} S[x_c]} \, . \tag{2.55}$$

Indeed, the stationarity of the action must result in a critical point (critical trajectory) $x_c(t)$ where Eq. (2.54) holds. Usually $x_c(t)$ is a minimum of the action $S[x(t)]$ but, in principle, $x_c(t)$ could be also a maximum or a saddle. Notice that $x_c(t)$ is obviously not extremizing the complex exponential function: the basic idea is that Eq. (2.55) gives, however, the main contribution to the quantum propagator $K(x_F, t_F | x_I, t_I)$ because, as written above, trajectories far from $x_c(t)$ somehow compensate destructively due to phase interference. In mathematics, this idea is called steepest-descent or saddle-point method and $x_c(t)$ is the saddle point of the complex exponential. Therefore, the name we give to Eq. (2.55) is saddle-point approximation. In high-energy physics, Eq. (2.55) is called tree-level approximation, while in condensed-matter physics it is indicated as a mean-field approximation. Eq. (2.55) is sometimes also denoted as lowest-order semiclassical approximation.

The next-to-leading order contribution consists in the integral of the quadratic expansion term of the action, since we expect, for stationarity, to have a null first-order contribution. Setting

$$x(t) = x_c(t) + \delta x(t) , \tag{2.56}$$

where $\delta x(t)$ is a small fluctuation around the classical solution $x_c(t)$, we have

$$S[x] = S[x_c] + \frac{1}{2} \int_{t_I}^{t_F} dt \int_{t_I}^{t_F} dt' \, \delta x(t) \, \frac{\delta^2 S[x_c]}{\delta x(t) \delta x(t')} \, \delta x(t') + \ldots \tag{2.57}$$

In the case of a nonrelativistic particle of mass m in the presence of the external potential $U(x)$ the second order functional derivative of the action functional is

$$\frac{\delta^2 S[x_c]}{\delta x(t) \delta x(t')} = \left(-m \frac{d^2}{dt^2} - \frac{\partial^2 U(x_c(t))}{\partial x^2} \right) \delta(t - t') , \tag{2.58}$$

that is a sort of harmonic problem under the effect of a time-dependent effective frequency

$$\omega_{\text{eff}}(x_c(t)) = \sqrt{\frac{1}{m} \frac{\partial^2 U(x_c(t))}{\partial x^2}} \tag{2.59}$$

which depends on the classical path $x_c(t)$.

By inserting Eq. (2.57) into Eq. (2.17), and noticing that the measure is invariant under shifts, so we can substitute $\mathcal{D}[\delta x]$ to $\mathcal{D}[x]$, we obtain

$$K(x_F, t_F | x_I, t_I) = e^{\frac{i}{\hbar} S[x_c]} \cdot \tag{2.60}$$
$$\int_{\delta x(t_I)=0}^{\delta x(t_F)=0} \mathcal{D}[\delta x] \exp \left[\frac{1}{2} \int_{t_I}^{t_F} dt \int_{t_I}^{t_F} dt' \, \delta x(t) \, \frac{i}{\hbar} \frac{\delta^2 S[x_c]}{\delta x(t) \delta x(t')} \, \delta x(t') \right]$$

at the Gaussian level. In this way we represent the propagator as a product of the saddle point solution and a Gaussian-like term due to fluctuations. The theory of Fresnel and Gaussian integrals provides us with the tool to evaluate this term. Taking into account Eq. (2.23) and remembering the definition of the functional integral as a limit of a multidimensional integral for the number of dimensions going to infinity, we have

$$\int_{\delta x(t_I)=0}^{\delta x(t_F)=0} \mathcal{D}[\delta x] \exp\left[\frac{1}{2}\int_{t_I}^{t_F} dt \int_{t_I}^{t_F} dt'\, \delta x(t)\, \frac{i}{\hbar}\, \frac{\delta^2 S[x_c]}{\delta x(t)\delta x(t')}\, \delta x(t')\right]$$
$$= \mathcal{C}\det\left[\frac{i}{\hbar}\frac{\delta^2 S[x_c]}{\delta x(t)\delta x(t')}\right]^{-1/2}, \tag{2.61}$$

where $\mathcal{C}$ is a constant that does not depend on the external potential $U(x)$. We will determine it in the next subsection. In general, in the continuum limit both $\mathcal{C}$ and $\det[\frac{i}{\hbar}\frac{\delta^2 S[x_c]}{\delta x(t)\delta x(t')}]^{1/2}$ are divergent while their ratio $F[x_c]$ is finite. Thus, we can write

$$K(x_F, t_F|x_I, t_I) = \mathcal{C}\det\left[\frac{i}{\hbar}\frac{\delta^2 S[x_c]}{\delta x(t)\delta x(t')}\right]^{-1/2} e^{\frac{i}{\hbar}S[x_c]} = F[x_c]\, e^{\frac{i}{\hbar}S[x_c]}\,, \tag{2.62}$$

that is called the stationary phase approximation, where

$$F[x_c] = \mathcal{C}\det\left[\frac{i}{\hbar}\frac{\delta^2 S[x_c]}{\delta x(t)\delta x(t')}\right]^{-1/2} \tag{2.63}$$

is the Gaussian correction to the quite raw saddle point approximation, Eq. (2.55). In high-energy physics, it is denoted as one-loop approximation, while in condensed-matter physics it is called as Gaussian approximation. Sometimes it is also indicated as first-order semiclassical approximation. The stationary phase approximation is a fundamental concept in asymptotic analysis, used for functions obtained by integrating with respect to a rapidly-changing complex exponential. This technique was created in the 19th century by George Stokes and William Thompson (Lord Kelvin). It is similar to Laplace's method and the method of steepest descent. In the context of path integral, it is also called Gaussian approximation but, more often, semiclassical approximation because one considers only quantum paths close to the classical one.

2.3.1 Particle in a Harmonic Potential

In the case of a particle of mass m in the harmonic potential

$$U(x) = \frac{m}{2}\omega^2 x^2 \tag{2.64}$$

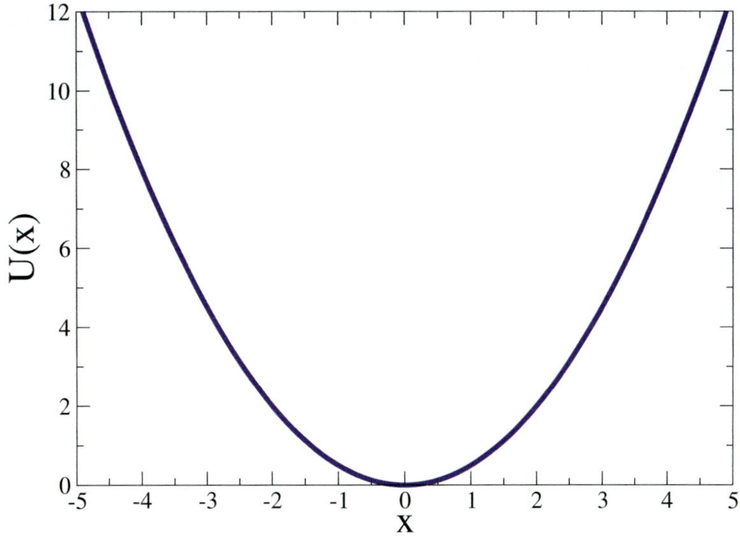

Figure 2.2
Harmonic potential $U(x) = m\omega^2 x^2/2$ setting $m\omega^2 = 1$.

with a constant frequency ω, the stationary phase is exact because the action

$$S[x] = \int_{t_I}^{t_F} dt \left(\frac{m}{2}\dot{x}(t)^2 - \frac{m}{2}\omega^2 x(t)^2 \right) \tag{2.65}$$

is quadratic with respect to $x(t)$ and the effective frequency of Eq. (2.59) is time-independent and coincides with ω. In Fig. 2.2 we plot the typical parabolic shape of the harmonic potential.

In this specific problem, it is straightforward to find the following classical path

$$x_c(t) = \frac{x_F \sin(\omega(t - t_I)) + x_I \sin(\omega(t_F - t))}{\sin(\omega(t_F - t_I))} \tag{2.66}$$

and from it the classical action

$$S[x_c] = \frac{m\omega}{2} \left[(x_F^2 + x_I^2)\cot(\omega(t_F - t_I)) - \frac{2x_F x_I}{\sin(\omega(t_F - t_I))} \right] . \tag{2.67}$$

Moreover, the second-order functional derivative of the action functional can be written as follows

$$\frac{i}{\hbar}\frac{\delta^2 S[x_c]}{\delta x(t)\delta x(t')} = \frac{i}{\hbar}m\left(-\frac{d^2}{dt^2} - \omega^2\right)\delta(t-t')\,, \tag{2.68}$$

where this differential operator satisfies the eigenvalue equation

$$\frac{i}{\hbar}m\left(-\frac{d^2}{dt^2} - \omega^2\right)\delta x_n(t) = \lambda_n\,\delta x_n(t) \tag{2.69}$$

with eigenvalues

$$\lambda_n = \frac{i}{\hbar}m\left[\left(\frac{\pi n}{(t_F - t_I)}\right)^2 - \omega^2\right] \tag{2.70}$$

and eigenfunctions

$$\delta x_n(t) = \sqrt{\frac{2}{(t_F - t_I)}}\sin\left(\frac{n\pi}{t_F - t_I}(t - t_I)\right) \tag{2.71}$$

subjected to the Dirichlet boundary conditions $\delta x_n(t_I) = \delta x_n(t_F) = 0$.

The product of all the eigenvalues λ_n is directly related to the determinant of the operator of Gaussian fluctuations:

$$\begin{aligned}\det\left[\frac{i}{\hbar}\frac{\delta^2 S[x_c]}{\delta x(t)\delta x(t')}\right] &= \det\left[\frac{i}{\hbar}m\left(-\frac{d^2}{dt^2} - \omega^2\right)\right] = \prod_{n=1}^{\infty}\lambda_n \\ &= \frac{i}{\hbar}m\prod_{n=1}^{\infty}\left[\left(\frac{\pi n}{(t_F - t_I)}\right)^2 - \omega^2\right].\end{aligned} \tag{2.72}$$

From Eq. (2.63) we have that

$$F_{\text{harm}}[x_c] = \mathcal{C}\left(\frac{i}{\hbar}m\prod_{n=1}^{\infty}\left[\left(\frac{\pi n}{(t_F - t_I)}\right)^2 - \omega^2\right]\right)^{-1/2}. \tag{2.73}$$

The normalization constant $\mathcal{C}$ is obtained by considering the propagator of the free particle ($\omega = 0$), for which we have the exact result, Eq. (2.48):

$$F_{\text{free}}[x_c] = \mathcal{C} \left(\frac{i}{\hbar} m \prod_{n=1}^{\infty} \left[\left(\frac{\pi n}{(t_F - t_I)} \right)^2 \right] \right)^{-1/2} = \sqrt{\frac{m}{2\pi i \hbar (t_F - t_I)}}\,, \tag{2.74}$$

namely

$$\mathcal{C} = \sqrt{\frac{m}{2\pi i \hbar (t_F - t_I)}} \sqrt{\frac{i}{\hbar} m \prod_{n=1}^{\infty} \left[\left(\frac{\pi n}{(t_F - t_I)} \right)^2 \right]}\,. \tag{2.75}$$

This is the constant that must be inserted into Eq. (2.73). It follows that

$$\begin{aligned} F_{\text{harm}}[x_c] &= \sqrt{\frac{m}{2\pi i \hbar (t_F - t_I)}} \sqrt{\prod_{n=1}^{\infty} \frac{\left(\frac{\pi n}{(t_F - t_I)} \right)^2}{\left(\frac{\pi n}{(t_F - t_I)} - \omega^2 \right)^2}} \\ &= \sqrt{\frac{m}{2\pi i \hbar (t_F - t_I)}} \sqrt{\prod_{n=1}^{\infty} \frac{1}{1 - \left(\frac{\omega (t_F - t_I)}{\pi n} \right)^2}} \\ &= \sqrt{\frac{m}{2\pi i \hbar (t_F - t_I)}} \sqrt{\frac{\omega (t_F - t_I)}{\sin(\omega (t_F - t_I))}}\,. \end{aligned} \tag{2.76}$$

Here we have used this exact representation of $\sin(x)/x$ in terms of an infinite product:

$$\frac{\sin(x)}{x} = \prod_{n=1}^{\infty} \left(1 - \frac{x^2}{\pi^2 n^2} \right)\,. \tag{2.77}$$

Thus, we finally find

$$F_{\text{harm}}[x_c] = \sqrt{\frac{m\omega}{2\pi i \hbar \sin(\omega (t_F - t_I))}}\,, \tag{2.78}$$

where, quite remarkably, this Gaussian correction depends only on the time interval $t_F - t_I$ and it is independent of the initial and final coordinates (x_I and x_F). In conclusion, the full propagator of the harmonic oscillator is given by

$$K_{\text{harm}}(x_F,t_F|x_I,t_I) = \sqrt{\frac{m\omega}{2\pi i\hbar\sin(\omega(t_F-t_I))}} \times e^{\frac{i}{\hbar}\frac{m\omega}{2}\left[(x_F^2+x_I^2)\cot(\omega(t_F-t_I))-\frac{2x_Fx_I}{\sin(\omega(t_F-t_I))}\right]} . \tag{2.79}$$

For the sake of completeness, we remind that having the propagator one immediately gets the time-dependent wavefunction. Thus, for a particle in a harmonic potential we obtain

$$\psi(x,t) = \int_{-\infty}^{+\infty} dx' \sqrt{\frac{m\omega}{2\pi i\hbar\sin(\omega t)}}\, e^{\frac{i}{\hbar}\frac{m\omega}{2}\left[(x^2+(x')^2)\cot(\omega t)-\frac{2\,x\,x'}{\sin(\omega t)}\right]}\,\psi_0(x') \tag{2.80}$$

with $\psi_0(x) = \psi(x,0)$ the initial wavefunction.

2.3.2 Van Vleck-Morette Formula

Within the stationary phase approximation, we have seen that

$$K(x_F,t_F|x_I,t_I) = \mathcal{C}\det\left[\frac{i}{\hbar}\frac{\delta^2 S[x_c]}{\delta x(t)\delta x(t')}\right]^{-1/2} e^{\frac{i}{\hbar}S[x_c]} . \tag{2.81}$$

It is possible to prove rigorously (Kleinert, 2009) that this expression can be rewritten as follows

$$K(x_F,t_F|x_I,t_I) = \frac{1}{\sqrt{2\pi i\hbar}}\sqrt{-\frac{\partial^2 S[x_c]}{\partial x_I\partial x_F}}\, e^{\frac{i}{\hbar}S[x_c]} . \tag{2.82}$$

In other words, the amplitude $F[x_c]$ of Gaussian fluctuations satisfies the equation

$$F[x_c] = \mathcal{C}\det\left[\frac{i}{\hbar}\frac{\delta^2 S[x_c]}{\delta x(t)\delta x(t')}\right]^{-1/2} = \frac{1}{\sqrt{2\pi i\hbar}}\sqrt{-\frac{\partial^2 S[x_c]}{\partial x_I\partial x_F}} . \tag{2.83}$$

This remarkable formula was found for the first time by John Hasbrouck Van Vleck in 1928, in the context of semiclassical quantization. Later it was generalized by Cecile Morette in the framework of Feynman's path integral. The formula involves the classical action $S[x_c(t)]$, that is a function of the initial x_I and final position x_F, i.e., $S[x_c(t)] = S[x_c](x_I, x_F)$, and its second partial derivative with respect to x_I and x_F. Remind that the second functional derivative, which appears in Eq. (2.81), is a quite different mathematical object with respect to the second partial derivative, which appears in Eq. (2.82). Let us verify that the Van Vleck-Morette formula reproduces the results, we have previously obtained for the free particle and the harmonic oscillator.

In the case of the free particle, the classical action is

$$S[x_c(t)] = \frac{m}{2}\frac{(x_F - x_I)^2}{(t_F - t_I)} . \tag{2.84}$$

If follows that

$$\frac{\partial S[x_c(t)]}{\partial x_I} = -m\frac{(x_F - x_I)}{(t_F - t_I)} \tag{2.85}$$

and also

$$\frac{\partial^2 S[x_c(t)]}{\partial x_I \partial_F} = -\frac{m}{(t_F - t_I)} . \tag{2.86}$$

Then we have

$$\frac{1}{\sqrt{2\pi i\hbar}}\sqrt{-\frac{\partial^2 S[x_c]}{\partial x_I \partial x_F}} = \frac{1}{\sqrt{2\pi i\hbar}}\sqrt{\frac{m}{(t_F - t_I)}} = \sqrt{\frac{m}{2\pi i\hbar(t_F - t_I)}} \tag{2.87}$$

that is indeed the expected result.

In the case of the harmonic oscillator, the classical action is instead

$$S[x_c] = \frac{m\omega}{2}\left[(x_F^2 + x_I^2)\cot(\omega(t_F - t_I)) - \frac{2x_F x_I}{\sin(\omega(t_F - t_I))}\right] . \tag{2.88}$$

If follows that

$$\frac{\partial S[x_c(t)]}{\partial x_I} = -m\omega\left[x_I \cot(\omega(t_F - t_I)) + \frac{x_F}{\sin(\omega(t_F - t_I))}\right] \tag{2.89}$$

and also

$$\frac{\partial^2 S[x_c(t)]}{\partial x_I \partial_F} = -\frac{m\omega}{\sin(\omega(t_F - t_I))} . \tag{2.90}$$

Then we have

$$\frac{1}{\sqrt{2\pi i\hbar}}\sqrt{-\frac{\partial^2 S[x_c]}{\partial x_I \partial x_F}} = \frac{1}{\sqrt{2\pi i\hbar}}\sqrt{\frac{m\omega}{\sin(\omega(t_F - t_I))}}$$
$$= \sqrt{\frac{m\omega}{2\pi i\hbar \sin(\omega(t_F - t_I))}} \qquad (2.91)$$

that is again the expected result.

2.3.3 Action on the Classical Trajectory

Apart from some very simple cases, the determination of the action $S[x(t)]$ on the classical trajectory $x_c(t)$ with boundary conditions $x_c(t_I) = x_I$ and $x_c(t_F) = x_F$ is not an easy task. The wanted classical action is formally given by

$$S[x_c] = \int_{t_F}^{t_F} L(x_c, \dot{x}_c)\, dt, \qquad (2.92)$$

where

$$L(x_c, \dot{x}_c) = \frac{m}{2}\dot{x}_c^2 - U(x_c) \qquad (2.93)$$

is the Lagrangian function with $\dot{x}_c(t)$ the velocity at time t. Here $x_c(t)$ must satisfy Newton's equation

$$m\ddot{x}_c = -\frac{\partial U}{\partial x}(x = x_c) \qquad (2.94)$$

and the energy

$$E_c = \frac{m}{2}\dot{x}_c^2 + U(x_c) \qquad (2.95)$$

is a constant of motion. Considering the linear momentum

$$p_c = m\dot{x}_c \qquad (2.96)$$

we can write

$$E_c = \frac{p_c^2}{2m} + U(x_c) \qquad (2.97)$$

where

$$H(x_c, p_c) = \frac{p_c^2}{2m} + U(x_c) \qquad (2.98)$$

is the Hamiltonian of the system calculated along the classical trajectory. From Eq. (2.97) we also have

$$p_c = \sqrt{2m(E_c - U(x_c))} \ . \tag{2.99}$$

Remembering the Legendre transformation between the Lagrangian L and the Hamiltonian H, namely

$$L = p_c \dot{x}_c - H(x_c, p_c) = p_c \dot{x}_c - E_c \ , \tag{2.100}$$

the action functional (2.93) can be rewritten as

$$S[x_c] = \int_{t_I}^{t_F} \left(p_c \frac{dx_c}{dt} - E_c \right) dt = \int_{x_I}^{x_F} p_c \, dx_c - E_c(t_F - t_I) \tag{2.101}$$

or better

$$S[x_c] = \int_{x_I}^{x_F} \sqrt{2m(E_c - U(x))} \, dx - E_c(t_F - t_I) \ . \tag{2.102}$$

However, this is not the final result because in our problem, E_c is not an independent variable: it depends on the initial data (x_I, t_I) and final data (x_F, t_I). To obtain an implicit formula for E_c, we notice that from Eq. (2.95) it follows

$$\sqrt{\frac{2}{m}(E_c - U(x_c))} = \frac{dx_c}{dt} \tag{2.103}$$

from which

$$\int_{t_I}^{t_F} dt = \int_{x_I}^{x_F} \frac{dx_c}{\sqrt{\frac{2}{m}(E_c - U(x_c))}} \tag{2.104}$$

and finally

$$(t_F - t_I) = \sqrt{\frac{m}{2}} \int_{x_I}^{x_F} \frac{dx}{\sqrt{E_c - U(x)}} \ . \tag{2.105}$$

Eqs. (2.102) and (2.105) must be solved self-consistently to get the action $S[x_c]$ as a function of x_I, x_F and $t_F - t_I$. For the sake of completeness, we observe that in this framework, the classical trajectory $x_c(t)$ can be evaluated by considering Eq. (2.105) with $t_F = t$ and $x_F = x_c(t)$. Thus, we have

$$t = t_I + \sqrt{\frac{m}{2}} \int_{x_I}^{x_c(t)} \frac{dx}{\sqrt{E_c - U(x)}} \qquad (2.106)$$

that is an implicit formula for $x_c(t)$, where E_c must be again derived with Eq. (2.105).

It is very important to underline that, as discussed in the first chapter, for any potential $U(x)$ rising faster than quadratically at large positive and negative x, there are many classical trajectories $x_c(t)$ which satisfy the boundary conditions $x_c(t_I) = x_I$ and $x_c(t_F) = x_F$. By increasing the initial velocity, the particle returns to the starting point within the same time interval $t_F - t_I$ after making several bounces forward and backward at the turning points. Thus, these trajectories are periodic orbits with the same closing time $t_F - t_I$, but after a different number of turns. These periodic orbits do not have the same period τ nor the same energy E_c, which changes as the initial speed changes. In this case the n-th energy $E_c^{(n)}$ is determined from the knowledge of the time interval $t_F - t_I$ and boundary conditions x_I and x_F by the formula

$$(t_F - t_I) = n \sqrt{\frac{m}{2}} \int_{x_I}^{x_F} \frac{dx}{\sqrt{E_c^{(n)} - U(x)}} \, , \qquad (2.107)$$

that is a generalization of Eq. (2.105) with the inclusion of the integer number $n = 1, 2, 3, ...$ Similarly, the n-th classical trajectory $x_c^{(n)}(t)$ is obtained from

$$t = t_I + n \sqrt{\frac{m}{2}} \int_{x_I}^{x_c^{(n)}(t)} \frac{dx}{\sqrt{E_c^{(n)} - U(x)}} \, . \qquad (2.108)$$

In this way, the corresponding n-th action is

$$S[x_c^{(n)}] = n \int_{x_I}^{x_F} \sqrt{2m(E_c^{(n)} - U(x))} \, dx - E_c^{(n)}(t_F - t_I) \, . \quad (2.109)$$

2.3.4 Particle in Free Fall

For a particle in free fall along the vertical x-axis under the effect of the gravitational potential

$$U(x) = -mgx \; , \tag{2.110}$$

that is a linear potential with $g = 9.81$ m/s^2, the gravitational acceleration, the Lagrangian is given by

$$L = \frac{m}{2}\dot{x}^2 + mgx \tag{2.111}$$

and the Euler-Lagrange equation for the classical trajectory $x_c(t)$ reads

$$\ddot{x}_c(t) = g \; . \tag{2.112}$$

Following the procedure illustrated in previous sections, it is not very difficult to find the action calculated on this classical trajectory. The results is

$$S[x_c] = \frac{m}{2}\left[\frac{(x_F - x_I)^2}{(t_F - t_I)} + g(x_I + x_F)(t_F - t_I) - \frac{g^2}{12}(t_F - t_I)^3\right] \tag{2.113}$$

As expected, by using the Van Vleck-Morette formula, one recognizes that the Gaussian correction is the same as the free particle. The difference is only in the action functional. In conclusion, we obtain

$$\begin{aligned} K_{\text{fall}}(x_F, t_F | x_I, t_I) \;\; &= \;\; \sqrt{\frac{m}{2\pi i\hbar(t_F - t_I)}} \\ &\quad \times e^{\frac{im}{2\hbar}\left[\frac{(x_F - x_I)^2}{(t_F - t_I)} + g(x_I + x_F)(t_F - t_I) - \frac{g^2}{12}(t_F - t_I)^3\right]} \end{aligned} \tag{2.114}$$

for the quantum propagator of a particle of mass m in free fall with g, the gravitational acceleration. This is an exact result because the confining potential is linear, and in this case, the stationary phase approximation is exact.

2.4 Green Functions

The time-dependent outgoing (or retarded) Green function $G(x,t|x',t')$ is defined through the propagator $K(x,t|x',t')$ as

$$G(x,t|x',t') = K(x,t|x',t')\,\Theta(t-t')\ , \tag{2.115}$$

where $\Theta(t)$ is the Heaviside step function, such that $d\Theta(t)/dt = \delta(t)$ is the Dirac delta function. Taking into account Eq. (2.9), one immediately deduces that the Green function satisfies the equation

$$\left(i\hbar\frac{\partial}{\partial t} + \frac{\hbar^2}{2m}\frac{\partial^2}{\partial x^2} - U(x)\right) G(x,t|x',t') = i\hbar\,\delta(t-t')\ , \tag{2.116}$$

whose formal solution is

$$G(x,t|x',t') = \frac{i\hbar\,\delta(t-t')}{i\hbar\frac{\partial}{\partial t} + \frac{\hbar^2}{2m}\frac{\partial^2}{\partial x^2} - U(x)}\ . \tag{2.117}$$

The n-point disconnected correlation function $G^{(n)}(t_1,...,t_n)$, also called n-point disconnected Green function, of the position operator $\hat{x}(t)$ with respect to the ground state $|GS\rangle$ of the Hamiltonian $\hat{H}$ of the system is defined as

$$G^{(n)}(t_1,...,t_n) = \langle GS|\hat{T}[\hat{x}(t_1)\ldots\hat{x}(t_n)]|GS\rangle\ , \tag{2.118}$$

where $\hat{T}$ is the time-ordered operator. For instance, in the 2-point case it is such that

$$\hat{T}[\hat{x}(t_1)\,\hat{x}(t_2)] = \begin{cases} x(t_1)\,x(t_2) & t_1 > t_2 \\ x(t_2)\,x(t_1) & t_2 > t_1 \end{cases} \tag{2.119}$$

Quite remarkably, the time ordering is encoded in the path integral formulation, and the n-point correlator can be written as follows

$$G^{(n)}(t_1,...,t_n) = \frac{\int_{x(-\infty)=x_I}^{x(+\infty)=x_F} \mathcal{D}[x(t)]\,x(t_1)\ldots x(t_n)\,e^{\frac{i}{\hbar}S[x(t)]}}{\int_{x(-\infty)=x_I}^{x(+\infty)=x_F} \mathcal{D}[x(t)]\,e^{\frac{i}{\hbar}S[x(t)]}}\ . \tag{2.120}$$

This relevant result is justified by the fact that the wildly oscillating sums of the path integral are dominated by the ground state $|GS\rangle$ in the limit of a large time and its phase contribution is removed by inserting the denominator in the previous formula. Clearly it follows that

$$G^{(0)} = \frac{\int_{x(-\infty)=x_I}^{x(+\infty)=x_F} \mathcal{D}[x(t)]\, e^{\frac{i}{\hbar}S[x(t)]}}{\int_{x(-\infty)=x_I}^{x(+\infty)=x_F} \mathcal{D}[x(t)]\, e^{\frac{i}{\hbar}S[x(t)]}} = K(x_F, +\infty | x_I, -\infty)\,. \tag{2.121}$$

To get a better understanding of these statements, let us consider the decomposition (2.11) of the propagator setting $t' = -\bar{t}/2$ and $t = \bar{t}/2$. We have

$$K\big(x, \frac{\bar{t}}{2} | x', -\frac{\bar{t}}{2}\big) = \sum_n \phi_n(x)\, \phi_n^*(x')\, e^{-\frac{i}{\hbar}E_n \bar{t}}\,. \tag{2.122}$$

The sum is indeed dominated by the ground-state wavefunction $\phi_0(x)$ in the limit of $\bar{t} \to +\infty$, i.e.

$$K\big(x, \frac{\bar{t}}{2} | x', -\frac{\bar{t}}{2}\big) \to \phi_0(x)\, \phi_0^*(x')\, e^{-\frac{i}{\hbar}E_0 \bar{t}}\,, \tag{2.123}$$

where $\phi_0(x) = \langle x | GS \rangle$.

We can introduce a functional $Z[J]$ encapsulating all the informations about the n-point correlation function. This is given by

$$Z[J] = \int_{x(-\infty)=x_I}^{x(+\infty)=x_F} \mathcal{D}[x(t)]\, e^{\frac{i}{\hbar}\left(S[x(t)] + \int_{-\infty}^{+\infty} dt J(t) x(t)\right)}\,, \tag{2.124}$$

where $J(t)$ is a complex source term. $Z[J]$ is called the generating functional because the functional differentiations with respect to $J(t)$ gives the correlation functions. In particular,

$$G^{(n)}(t_1, ..., t_n) = \left(\frac{\hbar}{i}\right)^n \frac{1}{Z[0]} \frac{\delta^2 Z[J]}{\delta J(t_1) \dots \dots \delta J(t_n)}\bigg|_{J=0}\,. \tag{2.125}$$

This expression can be rewritten as

$$G^{(n)}(t_1, ..., t_n) = \left(\frac{\hbar}{i}\right)^{n-1} \frac{\delta^n Z[J]}{\delta J(t_1) \dots \dots \delta J(t_n)}\bigg|_{J=0} \tag{2.126}$$

having introduced the associated functional $W[J]$ such that

$$Z[J] = e^{\frac{i}{\hbar}W[J]} \tag{2.127}$$

or equivalently

$$W[J] = -\frac{i}{\hbar}\ln(Z[J]) \; . \tag{2.128}$$

In the last part of this chapter, we will discuss the thermodynamic partition function $\mathcal{Z}$, that is, the imaginary-time version of the real-time functional $Z[0]$. Similarly, we will consider the Helmholtz free energy F, that is related to the generating functional $W[0]$.

2.4.1 Fixed-Energy Green function

The fixed-energy Green function, sometimes called fixed-energy propagator, is defined as

$$\begin{aligned} G_E(x_F|x_I) &= \int_{-\infty}^{+\infty} G(x_F,t|x_I,0)\, e^{\frac{i}{\hbar}E\,t}\, dt \\ &= \int_0^{+\infty} K(x,t|x',0)\, e^{\frac{i}{\hbar}E\,t}\, dt \end{aligned} \tag{2.129}$$

remembering that $G(x_F,t|x_I,0) = K(x_F,t|x_I,0)\,\Theta(t)$ with $\Theta(t)$ The Heaviside step function. Adopting the semiclassical saddle point and stationary phase approximation, Eq. (2.62), we have

$$G_E(x_F|x_I) \simeq \int_0^{+\infty} F[x_c]\, e^{\frac{i}{\hbar}S[x_c]}\, e^{\frac{i}{\hbar}E\,t}\, dt \; . \tag{2.130}$$

Performing a further saddle point approximation, we get

$$G_E(x_F|x_I) \simeq F_E[x_c]\, e^{\frac{i}{\hbar}(S_E[x_c]+Et_s(E))} \; , \tag{2.131}$$

where $t_s = t_s(E)$ is the stationary time, given by

$$\frac{\partial S[x_c]}{\partial t}|_{t=t_s} + E = 0 \; . \tag{2.132}$$

Here the action $S_E[x_c]$ and the amplitude $F_E[x_c]$ are calculated along the classical trajectory at fixed energy E. Explicitly, taking into account Eq. (2.105), we find

$$t_s(E) = \sqrt{\frac{m}{2}} \int_{x_I}^{x_F} \frac{dx}{\sqrt{E - U(x)}} \,. \tag{2.133}$$

In conclusion, we have

$$G_E(x_F|x_I) \simeq F_E[x_c]\, e^{\frac{i}{\hbar}\int_{x_I}^{x_F} \sqrt{2m(E-U(x))}} \,, \tag{2.134}$$

where $F_E[x_c]$ can be written as

$$F_E[x_c] = \frac{1}{\sqrt{2\pi i\hbar}} \sqrt{-\frac{\partial^2 S_E[x_c]}{\partial x_I \partial x_F}} \,. \tag{2.135}$$

invoking the Van Vleck-Morette formula, Eq. (2.83), and considering that

$$S_E[x_c] = \int_{x_I}^{x_F} \sqrt{2m(E - U(x))}dx - mE \int_{x_I}^{x_F} \frac{dx}{\sqrt{2m(E - U(x))}} \,. \tag{2.136}$$

2.5 Propagator of Quantum States

Let us consider $|I\rangle$ a generic initial state of the particle and $|F\rangle$ a generic final state of the particle. The propagator $K(F, t_F|I, t_I)$ of the particle to move from the quantum state $|I\rangle$ at time t_I to the quantum state $|F\rangle$ at time t_F is given by

$$K(F, t_F|I, t_I) = \langle F|e^{-\frac{i}{\hbar}\hat{H}(t_F - t_I)}|I\rangle \,. \tag{2.137}$$

Taking into account the coordinate resolution of the identity

$$\hat{1} = \int_{-\infty}^{+\infty} dx|x\rangle\langle x| \tag{2.138}$$

we get

$$
\begin{aligned}
&\langle F|e^{-\frac{i}{\hbar}\hat{H}(t_F-t_I)}|I\rangle \\
&\quad = \langle F|\hat{1}\, e^{-\frac{i}{\hbar}\hat{H}(t_F-t_I)}\, \hat{1}|I\rangle \qquad (2.139)\\
&\quad = \langle F| \int_{-\infty}^{+\infty} dx_F |x_F\rangle\langle x_F|e^{-\frac{i}{\hbar}\hat{H}(t_F-t_I)} \int_{-\infty}^{+\infty} dx_I |x_I\rangle\langle x_I|I\rangle \\
&\quad = \int_{-\infty}^{+\infty} dx_F \int_{-\infty}^{+\infty} dx_I\, \langle F|x_F\rangle\langle x_F|e^{-\frac{i}{\hbar}\hat{H}(t_F-t_I)}|x_I\rangle\langle x_I|I\rangle \\
&\quad = \int_{-\infty}^{+\infty} dx_F \int_{-\infty}^{+\infty} dx_I\, F^*(x_F)\, I(x_I)\, \langle x_F|e^{-\frac{i}{\hbar}\hat{H}(t_F-t_I)}|x_I\rangle \, .
\end{aligned}
$$

Thus, in conclusion, we obtain the convolution formula

$$
K(F, t_F|I, t_I) = \int_{-\infty}^{+\infty} dx_F \int_{-\infty}^{+\infty} dx_I\, F^*(x_F)\, I(x_I)\, K(x_F, t_F|x_I, t_I) \, , \qquad (2.140)
$$

where

$$
I(x) = \langle x|I\rangle \qquad (2.141)
$$

is the wavefunction, i.e., the probability amplitude, of the particle in the state $|I\rangle$ at the position $|x\rangle$, and

$$
F^*(x) = \langle F|x\rangle \, , \qquad (2.142)
$$

is the conjugate wavefunction of the particle in the state $|F\rangle$ at the position $|x\rangle$.

Inserting Eq. (2.17) into Eq. (2.140), we find

$$
\begin{aligned}
K(F, t_F|I, t_I) \;&=\; \int_{-\infty}^{+\infty} dx_F \int_{-\infty}^{+\infty} dx_I\, F^*(x_F)\, I(x_I) \\
&\quad \times \int_{x(t_I)=x_I}^{x(t_F)=x_F} \mathcal{D}[x(t)]\, e^{\frac{i}{\hbar}S[x(t)]} \, . \qquad (2.143)
\end{aligned}
$$

This formula gives the propagator of quantum states in terms of the Feynman path integral and involves convolutions with respect to the wavefunctions of the two quantum states.

2.5.1 Propagator for the Density Matrix

For an initial quantum state $|I\rangle$, the corresponding density operator reads

$$\hat{\rho}_I = |I\rangle\langle I| \, . \tag{2.144}$$

Having the density operator $\hat{\rho}_I$, one can write its density matrix in the coordinate representation:

$$\rho_I(x,x') = \langle x|\hat{\rho}_I|x'\rangle = \langle x|I\rangle \, \langle I|x'\rangle = I(x)\, I^*(x') \, . \tag{2.145}$$

The density matrix $\rho(x,x',t)$ at time t can then be derived as

$$\rho(x,x',t) = \langle x|e^{-\frac{i}{\hbar}\hat{H}(t-t_I)}\hat{\rho}_I e^{\frac{i}{\hbar}\hat{H}(t-t_I)}|x'\rangle \, . \tag{2.146}$$

Explicitly we have

$$\begin{aligned}
&\rho(x,x',t) \\
&= \langle x|e^{-\frac{i}{\hbar}\hat{H}(t-t_I)}|I\rangle \, \langle I|e^{\frac{i}{\hbar}\hat{H}(t-t_I)}|x'\rangle \qquad (2.147)\\
&= \langle x|e^{-\frac{i}{\hbar}\hat{H}(t-t_I)} \int_{-\infty}^{+\infty} dx_I \, |x_I\rangle\langle x_I|I\rangle \, \langle I| \\
&\qquad \times \int_{-\infty}^{+\infty} dx'_I|x'_I\rangle\langle x'_I|e^{\frac{i}{\hbar}\hat{H}(t-t_I)}|x'\rangle \\
&= \int_{-\infty}^{+\infty} dx_I \int_{-\infty}^{+\infty} dx'_I \, \langle x|e^{-\frac{i}{\hbar}\hat{H}(t-t_I)}|x_I\rangle \, \langle x'_I|e^{\frac{i}{\hbar}\hat{H}(t-t_I)}|x'\rangle \, \langle x_I|I\rangle\langle I|x'_I\rangle \\
&= \int_{-\infty}^{+\infty} dx_I \int_{-\infty}^{+\infty} dx'_I \, \langle x|e^{-\frac{i}{\hbar}\hat{H}(t-t_I)}|x_I\rangle \, \langle x'_I|e^{\frac{i}{\hbar}\hat{H}(t-t_I)}|x'\rangle \, \rho_I(x_I,x'_I) \, .
\end{aligned}$$

Remembering the representation of the propagator in terms of the path integral, we get

$$\rho(x,x',t) = \int_{-\infty}^{+\infty} dx_I \int_{-\infty}^{+\infty} dx'_I \, K(x,t|x_I,t_I)\, K(x'_I,t_I|x',t)\, \rho_I(x_I,x'_I) \tag{2.148}$$

namely

$$\begin{aligned}
\rho(x,x',t) \;=\; & \int_{\mathbb{R}^2} dx_I dx'_I \int_{y(t_I)=x_I}^{y(t)=x} \mathcal{D}[y(\bar{t})] \int_{\tilde{y}(t_I)=x'_I}^{\tilde{y}(t)=x'} \mathcal{D}[\tilde{y}(\bar{t})] \\
& \times e^{\frac{i}{\hbar}(S[y(\bar{t})]-S[\tilde{y}(\bar{t})])} \, \rho_I(x_I,x'_I) \, .
\end{aligned} \tag{2.149}$$

Thus, we have obtained a double path integral: there is a doubling of the degrees of freedom, where the two independent paths $y(\bar{t})$ and $\tilde{y}(\bar{t})$ account respectively for the forward and backward evolution in time $\bar{t}$.

Eq. (2.149) is quite useful because it can be generalized to the case of a non-pure initial quantum state, and also to the case in which the system is not closed but instead coupled to a reservoir. For instance, an initial mixed quantum state is described by a slightly more sophisticated density operator $\hat{\rho}_I$, namely

$$\hat{\rho}_I = \sum_\alpha p_\alpha |I_\alpha\rangle\langle I_\alpha| \;, \tag{2.150}$$

where p_α is the probability of finding the pure quantum state $|I_\alpha\rangle$ in the mixed quantum state, with

$$\sum_\alpha p_\alpha = 1 \;. \tag{2.151}$$

A typical example of a mixed state is a thermal mixed state such that

$$p_\alpha = \frac{1}{\mathcal{Z}_1} e^{-\beta E_\alpha} \;, \tag{2.152}$$

where $\beta = 1/(k_B T)$ with k_B the boltzmann constant and T the temperature, E_α are the energy levels, and

$$\mathcal{Z}_1 = \sum_\alpha e^{-\beta E_\alpha} \tag{2.153}$$

is the partition function. The corresponding density matrix in the coordinate representation reads

$$\begin{aligned} \rho_I(x,x') &= \langle x|\hat{\rho}_I|x'\rangle = \langle x \sum_\alpha p_\alpha |I_\alpha\rangle \, \langle I_\alpha|x'\rangle = \sum_\alpha p_\alpha \langle x|I_\alpha\rangle \, \langle I_\alpha|x'\rangle \\ &= \sum_\alpha p_\alpha I_\alpha(x) \, I^*_\alpha(x') \;. \end{aligned} \tag{2.154}$$

Also in this case the density matrix $\rho(x,x',t)$ at time t can be then obtained as

$$\begin{aligned} \rho(x,x',t) &= \langle x|e^{-\frac{i}{\hbar}\hat{H}(t-t_I)}\hat{\rho}_I e^{\frac{i}{\hbar}\hat{H}(t-t_I)}|x'\rangle \\ &= \sum_\alpha p_\alpha \langle x|e^{-\frac{i}{\hbar}\hat{H}(t-t_I)}|I_\alpha\rangle \, \langle I_\alpha|e^{\frac{i}{\hbar}\hat{H}(t-t_I)}|x'\rangle \end{aligned} \tag{2.155}$$

It is then straightforward to prove that Eq. (2.149) holds also in this context of the mixed state, with $\rho_I(x, x')$ given by Eq. (2.154).

In the specific case of a thermal mixed state at equilibrium, i.e. such that Eq. (2.152) holds but also

$$\hat{H}|I_\alpha\rangle = E_\alpha |I_\alpha\rangle \ , \tag{2.156}$$

one immediately finds

$$\rho(x, x', t) = \frac{1}{\mathcal{Z}_1} \sum_\alpha e^{-\beta E_\alpha} I_\alpha(x)\, I_\alpha^*(x') = \rho_I(x, x') \ . \tag{2.157}$$

This means that, in this specific case, the density matrix is time independent. Moreover, we find

$$\mathcal{Z}_1 = \int dx\, \rho_I(x, x) \ , \tag{2.158}$$

which is a simple formula connecting the partition function $\mathcal{Z}_1$ to the diagonal part $\rho_I(x, x)$ of the density matrix at thermal equilibrium.

2.6 Thermodynamics and Wick Rotation

In the first chapter we have seen that in the canonical ensemble the partition function $\mathcal{Z}_N$ of a quantum system with Hamiltonian $\hat{H}$ at temperature T and N identical particles is given by

$$\mathcal{Z}_N = \mathrm{Tr}\left[e^{-\beta \hat{H}}\right] \tag{2.159}$$

where $Tr[\hat{A}]$ is the trace of the operator $\hat{A}$, i.e., the sum of all the diagonal elements of $\hat{A}$, and $\beta = 1/(k_B T)$ with k_B the Boltzmann constant. We have also seen that the thermal average of a generic quantum observable $\hat{A}$ is then defined as

$$\langle \hat{A} \rangle_T = \frac{1}{\mathcal{Z}_N} \mathrm{Tr}[\hat{A}\, e^{-\beta \hat{H}}] \ . \tag{2.160}$$

It follows that the internal energy $E = \langle \hat{H} \rangle_T$ is obtained directly from the relation

$$E = -\frac{1}{\beta}\frac{\partial}{\partial \beta} \ln(\mathcal{Z}_N) \tag{2.161}$$

taking into account Eqs. (2.159) and (2.160). It is useful to remind here, in the canonical ensemble, the free energy F of the system is related to the partition function by the formula

$$\mathcal{Z}_N = e^{-\beta F} , \tag{2.162}$$

or equivalently

$$F = -\frac{1}{\beta} \ln(\mathcal{Z}_N) . \tag{2.163}$$

The free energy F is related to the internal energy E and the entropy $\mathcal{S}$ by the thermodynamic formula

$$F = E - T\mathcal{S} . \tag{2.164}$$

Thus, knowing E and F, the entropy reads

$$\mathcal{S} = \frac{E - F}{T} . \tag{2.165}$$

However, the entropy $\mathcal{S}$ can be also derived directly from the free energy F as

$$\mathcal{S} = -\frac{\partial F}{\partial T} , \tag{2.166}$$

and consequently from the partition function $\mathcal{Z}_N$ as

$$\mathcal{S} = k_B \left(1 + T\frac{\partial}{\partial T}\right) \ln(\mathcal{Z}_N) . \tag{2.167}$$

The operator $e^{-\beta\hat{H}}$ which appears in Eqs. (2.159) and (2.160) is similar, but not equal, to the unitary operator $\hat{U}(t) = e^{-i\hat{H}/\hbar}$ of the time evolution. In particular,

$$e^{-\beta\hat{H}} = \hat{U}(t = -i\hbar\beta) . \tag{2.168}$$

In the simple case of a particle described by the Hamiltonian $\hat{H} = \hat{p}^2/(2m) + U(\hat{x})$, we can write

$$\mathcal{Z}_1 = \int_{-\infty}^{+\infty} dx_I \, \langle x_I | e^{-\beta\hat{H}} | x_I \rangle \tag{2.169}$$

and also

$$\begin{aligned} \langle x_I | e^{-\beta \hat{H}} | x_I \rangle &= \langle x_I | \hat{U}(t = -i\hbar\beta) | x_I \rangle \\ &= K(x_F = x_I, t_F = -i\hbar\beta | x_I, t_I = 0) \end{aligned} \quad (2.170)$$

considering the Feynman propagator $K(x_F, t_F | x_F, t_I) = \langle x_F | \hat{U}(t_F - t_I) | x_F \rangle$ with $x_F = x_I$, $t_I = 0$, and $t_F = -i\hbar\beta$. Thus, we have

$$\mathcal{Z}_1 = \int_{-\infty}^{+\infty} dx_I \, K(x_I, -i\hbar\beta | x_I, 0) \; . \quad (2.171)$$

Please note the direct link between these results and Eqs. (2.157) and (2.158) of the previous section. Indeed, $\langle x_I | e^{-\beta \hat{H}} | x_I \rangle$ is nothing else than the diagonal part of the density matrix at thermal equilibrium.

To make a deep connection with the path integral representation of the propagator, it is useful to introduce the so-called imaginary time $\tau \in [0, \hbar\beta]$, that is actually a real, non-negative number, such that

$$t = -i\tau \; . \quad (2.172)$$

This formula, proposed by Gian Carlo Wick in 1954, is called Wick rotation because it corresponds to a rotation of $\pi/2$ degrees for the complexified time t. In this way one obtains

$$\mathcal{Z}_1 = \int_{-\infty}^{+\infty} dx_I \int_{x(0)=x_I}^{x(\hbar\beta)=x_I} \mathcal{D}[x] \exp\left[-\frac{1}{\hbar} \int_0^{\hbar\beta} d\tau \left(\frac{m}{2} \dot{x}^2 + U(x) \right) \right] , \quad (2.173)$$

where each path $x(\tau)$ must satisfy the condition

$$x(0) = x(\hbar\beta) \; , \quad (2.174)$$

i.e., $x_I = x_F$. This condition can be enforced by expanding $x(\tau)$ into a Fourier series

$$x(\tau) = \sum_{n=-\infty}^{+\infty} x_n \, e^{-i\omega_n \tau} \; , \quad (2.175)$$

where $x_n^* = x_{-n}$ to ensure that $x(\tau)$ is real. The frequencies ω_n, called bosonic Matsubara frequencies, are not arbitrary but they are instead quantized

$$\omega_n = \frac{2\pi}{\hbar\beta} n = \frac{2\pi k_B T}{\hbar} n \tag{2.176}$$

ensuring the condition (2.174). When considered as functions on the entire τ axis, the paths $x(\tau)$ are periodic with period $\hbar\beta$, i.e.

$$x(\tau) = x(\tau + \hbar\beta) \ . \tag{2.177}$$

Notice that in the exponent of the path integral of Eq. (2.173) it appears the Euclidean action

$$S_E[x] = \int_0^{\hbar\beta} d\tau \left(\frac{m}{2} \dot{x}(\tau)^2 + U(x(\tau)) \right) \ . \tag{2.178}$$

Usually, for the sake of compactness, the integral with respect to the initial (and final) condition x_I is not written, and Eq. (2.173) is abbreviated as follows

$$\mathcal{Z}_1 = \int_{x(0)=x(\hbar\beta)} \mathcal{D}[x] \, e^{-S_E[x]/\hbar} \tag{2.179}$$

or, sometimes, as

$$\mathcal{Z}_1 = \oint \mathcal{D}[x] \, e^{-S_E[x]/\hbar} \ . \tag{2.180}$$

Explicitly, this Euclidean path integral means

$$\mathcal{Z}_1 = \int_{\mathbb{R}^N} \left(\prod_{n=1}^{N} dx_n \right) \left(\frac{m}{2\pi\hbar\Delta\tau} \right)^{N/2} e^{-\frac{1}{\hbar} \sum_{j=0}^{N-1} [\frac{m}{2} \frac{(x_{j+1}-x_j)^2}{\Delta\tau^2} + U(x_j)]\Delta\tau} \tag{2.181}$$

for $N \to \infty$, with $x_0 = x_N$ and $\Delta\tau = \hbar\beta/N$.

It is important to stress that the thermal average of a quantum observable $\hat{A} = A[\hat{x}]$ reads

$$\langle \hat{A} \rangle_T = \frac{1}{\mathcal{Z}_1} \int_{x(0)=x(\hbar\beta)} \mathcal{D}[x] \, A[x] \, e^{-S_E[x]/\hbar} \tag{2.182}$$

with $A[x]$ the classical analog of $A[\hat{x}]$.

One expects that in the zero-temperature limit, i.e., for $T \to 0$ or $\beta \to +\infty$, the thermal average corresponds to the average with respect to the ground state $|GS\rangle$. This is indeed the case:

$$\langle GS|\hat{A}|GS\rangle = \lim_{T\to 0}\langle \hat{A}\rangle_T = \frac{\int_{x(0)=x(+\infty)} \mathcal{D}[x]\, A[x]\; e^{-\frac{1}{\hbar}\int_0^{+\infty} d\tau\left(\frac{m}{2}\dot{x}^2+U(x)\right)}}{\int_{x(0)=x(+\infty)} \mathcal{D}[x] e^{-\frac{1}{\hbar}\int_0^{+\infty} d\tau\left(\frac{m}{2}\dot{x}^2+U(x)\right)}}, \tag{2.183}$$

under the assumption that $\hat{A}$ does not depend on two different time instants. Moreover, taking into account that

$$\mathcal{Z}_1 \to e^{\beta E_{gs}} \qquad \text{for} \qquad \beta \to +\infty \tag{2.184}$$

with E_{gs} the ground-state energy of the system, we have

$$E_{gs} = -\lim_{T\to 0} k_B T \ln(\mathcal{Z}_1)\ . \tag{2.185}$$

2.6.1 Free Particle at Thermal Equilibrium

From Eqs. (2.48) and (2.171), it is straightforward to derive the partition function of a free particle, taking into account that

$$K_{\text{free}}(x_I, -i\hbar\beta|x_I, 0) = \sqrt{\frac{m}{2\pi\hbar^2\beta}}\ . \tag{2.186}$$

It follows immediately

$$\mathcal{Z}_1 = \int_{\infty}^{+\infty} dx_I \sqrt{\frac{mk_BT}{2\pi\hbar^2}}\ . \tag{2.187}$$

This quantity is divergent because we are considering the infinite space interval $]-\infty, +\infty[$. Working with a finite interval of size L, for instance, $[0, L]$, we have

$$\mathcal{Z}_1 = \int_0^L dx_I \sqrt{\frac{mk_BT}{2\pi\hbar^2}} = \sqrt{\frac{mk_BT}{2\pi\hbar^2}}\, L\ . \tag{2.188}$$

Notice that the same result can be also obtained from

$$\mathcal{Z}_1 = \int_0^L dx_I \int_{-\infty}^{+\infty} \frac{dp}{2\pi\hbar}\, e^{-\beta\frac{p^2}{2m}}\ , \tag{2.189}$$

that is the familiar expression of the canonical partition function of the free particle in the phase space.

Heuristically, we can extend Eq. (2.188) to the more realistic three-dimensional problem of a particle confined in a volume $V = L^3$:

$$\mathcal{Z}_1 = \left(\frac{mk_BT}{2\pi\hbar^2}\right)^{3/2} V = \left(\frac{mk_BTV^{2/3}}{2\pi\hbar^2}\right)^{3/2} . \tag{2.190}$$

The free energy is then given by

$$F = -k_BT\ln(\mathcal{Z}_1) = -\frac{3}{2}\,k_BT\,\ln\left(\frac{mk_BTV^{2/3}}{2\pi\hbar^2}\right) \tag{2.191}$$

while the entropy reads

$$\mathcal{S} = -\left(\frac{\partial F}{\partial T}\right)_V = \frac{3}{2}\,k_B\,\ln\left(\frac{mk_BTV^{2/3}}{2\pi\hbar^2}\right) + \frac{3}{2}k_B\,. \tag{2.192}$$

Thus, the internal energy follows as

$$E = F + T\mathcal{S} = \frac{3}{2}k_BT\,, \tag{2.193}$$

that is the familiar formula of the thermal kinetic energy of one particle in three spatial dimensions.

2.6.2 Harmonic Oscillator at Thermal Equilibrium

The partition function of the harmonic oscillator can be derived from the propagator of Eq (2.79) by using Eq. (2.171). In particular, we find

$$\begin{aligned}
&K_{\text{harm}}(x_I, -i\hbar\beta|x_I, 0)\\
&= \sqrt{\frac{m\omega}{2\pi i\hbar\sin(-i\beta\hbar\omega)}}\,e^{\frac{i}{\hbar}\frac{m\omega}{2}\left[2x_I^2\cot(-i\beta\hbar\omega)-\frac{2x_I^2}{\sin(-i\beta\hbar\omega)}\right]}\\
&= \sqrt{\frac{m\omega}{2\pi\hbar\sinh(\beta\hbar\omega)}}\,e^{-\frac{m\omega}{\hbar}\left[\coth(\beta\hbar\omega)-\frac{1}{\sinh(\beta\hbar\omega)}\right]x_I^2}\\
&= \sqrt{\frac{m\omega}{2\pi\hbar\sinh(\beta\hbar\omega)}}\,e^{-\frac{m\omega}{\hbar}\frac{\cosh(\beta\hbar\omega)-1}{\sinh(\beta\hbar\omega)}x_I^2}
\end{aligned} \tag{2.194}$$

because $\sin(ix) = i\,\sinh(x)$ with $\sinh(x) = (e^x - e^{-x})/2$ the hyperbolic sine while $\cos(ix) = \cosh(x) = (e^x + e^{-x})/(2i)$ is the hyperbolic cosine. Then we have

$$\begin{aligned}
\mathcal{Z}_1 &= \int_{-\infty}^{+\infty} dx_I\, K_{\text{harm}}(x_I, -i\hbar\beta|x_I, 0) \\
&= \sqrt{\frac{m\omega}{2\pi\hbar\sinh(\beta\hbar\omega)}} \int_{-\infty}^{+\infty} dx_I\, e^{-\frac{m\omega}{\hbar}\frac{\cosh(\beta\hbar\omega)-1}{\sinh(\beta\hbar\omega)} x_I^2} \\
&= \frac{1}{\sqrt{2\left(\cosh(\beta\hbar\omega) - 1\right)}} = \frac{1}{\sqrt{4\sinh^2(\beta\hbar\omega/2)}} \\
&= \frac{1}{2\sinh(\beta\hbar\omega/2)} = \frac{e^{-\beta\hbar\omega/2}}{1 - e^{-\beta\hbar\omega}} \,.
\end{aligned} \tag{2.195}$$

The free energy

$$F = -k_B T \ln(\mathcal{Z}_1) = \frac{\hbar\omega}{2} + k_B T\, \ln\left(1 - e^{-\hbar\omega/(k_B T)}\right) \tag{2.196}$$

shows explicity the zero-point energy of the ground state. The entropy is instead

$$\mathcal{S} = -\left(\frac{\partial F}{\partial T}\right) = k_B \left[\frac{\hbar\omega}{k_B T} \frac{1}{e^{\hbar\omega/(k_B T)} - 1} - \ln\left(1 - e^{-\hbar\omega/(k_B T)}\right)\right] . \tag{2.197}$$

Finally, we get the internal energy

$$E = F + T\mathcal{S} = \frac{\hbar\omega}{2} + \frac{\hbar\omega}{e^{\hbar\omega/(k_B T)} - 1} = \frac{\hbar\omega}{2}\coth\left(\frac{\hbar\omega}{2k_B T}\right) , \tag{2.198}$$

where $\coth(x) = (e^x + e^{-x})/(e^x - e^{-x})$ is the hyperbolic cotangent. As expected, the internal energy contains both the zero-point energy and the energy of the single-mode Bose-Einstein thermal distribution.

2.6.3 Semiclassical Thermodynamics

In the high-temperature regime, i.e., with T very large and $\beta = 1/(k_B T)$ very small, the interval $[0, \beta\hbar]$ of integration in the Euclidean action of Eq. (2.173) is very short and $U(x(\tau)) \simeq$

$U(x(0)) = U(x_I)$. Consequently, using Eq. (2.188), the partition function reads

$$\mathcal{Z}_1 \simeq \int_{-\infty}^{+\infty} dx_I \, e^{-\beta U(x_I)} \int_{x(0)=x_I}^{x(\hbar\beta)=x_I} \mathcal{D}[x] \exp\left[-\frac{1}{\hbar}\int_0^{\hbar\beta} \mathrm{d}\tau \left(\frac{m}{2}\dot{x}^2\right)\right]$$
$$= \sqrt{\frac{mk_BT}{2\pi\hbar^2}} \int_{-\infty}^{+\infty} dx_I \, e^{-\beta U(x_I)} \,, \tag{2.199}$$

that is a quite useful high-temperature semiclassical formula. This result can be perturbatively improved in powers of $\hbar$. Including the next-to-leading correction (Zinn-Justin, 2021) one finds

$$\mathcal{Z}_1 = \sqrt{\frac{mk_BT}{2\pi\hbar^2}} \int_{-\infty}^{+\infty} dx_I \exp\left(-\beta U(x_I) - \beta^2 \frac{\hbar^2}{24m} U''(x_I) + O(\hbar^4)\right). \tag{2.200}$$

Thus, the particle is under the action of the effective potential

$$U_{\text{eff}}(x) = U(x) + \frac{\hbar^2}{24mk_BT} U''(x) + ... \tag{2.201}$$

which depends explicitly on the temperature and the reduced Planck constant.

A different semiclassical method is obtained by using the imaginary-time saddle-point approximation, which gives

$$\mathcal{Z}_1 \simeq \int_{-\infty}^{+\infty} dx_I \, e^{-\frac{1}{\hbar}\int_0^{\hbar\beta} d\tau\left(\frac{m}{2}\dot{x}_c(\tau)^2 + U(x_c(\tau))\right)} \tag{2.202}$$

where $x_c(\tau)$ is the classical trajectory such that

$$m\frac{d^2}{d\tau^2} x_c(\tau) = \frac{\partial U}{\partial x}(x = x_c(\tau)) \tag{2.203}$$

is the imaginary-time Newton's equation. Note that, in practice, it corresponds to changing the sign of the potential energy $U(x) \to -U(x)$. Eq. (2.203) must be equipped with the boundary condition

$$x_c(0) = x_c(\hbar\beta) = x_I \, . \tag{2.204}$$

Clearly, Eq. (2.202) can be improved by including Gaussian fluctuations in the framework of the imaginary-time stationary-phase approximation.

2.6.4 Many-Body Problem at Thermal Equilibrium

In the previous chapter we have seen that in the case of N identical particles of mass m confined by the trapping potential $U(\mathbf{r})$ and interacting with the interaction potential $V(\mathbf{r},\mathbf{r}')$, the quantum Hamiltonian is given by

$$\hat{H} = \sum_{j=1}^{N} \left(\frac{\hat{p}_j^2}{2m} + U(\hat{\mathbf{r}}_j) \right) + \frac{1}{2} \sum_{\substack{j,k=1 \\ j \neq k}}^{N} V(\hat{\mathbf{r}}_j, \hat{\mathbf{r}}_k) \,. \tag{2.205}$$

In the canonical ensemble the partition function $\mathcal{Z}_N$ of the system must take into account the fact that quantum identical particles are indistinguishable, and it reads

$$\mathcal{Z}_N = Tr[e^{-\beta\hat{H}}] = \frac{1}{N!} \sum_{\mathcal{P}} \xi^{\mathcal{P}} \langle \mathbf{r}_1\, \mathbf{r}_2 \ldots \mathbf{r}_N | e^{-\beta\hat{H}} | \mathbf{r}_{\mathcal{P}1}\, \mathbf{r}_{\mathcal{P}2} \ldots \mathbf{r}_{\mathcal{P}N} \rangle \,, \tag{2.206}$$

where $\mathcal{P}$ is one of the $N!$ permutations of the indices of the N position vectors of the particles. Here $\xi = 1$ for bosons and $\xi = -1$ for fermions because identical bosons must be symmetric with respect the exchange of two coordinates while fermions must be antisymmetric. In fact, the symbol $1^{\mathcal{P}}$ means 1 while the symbol $(-1)^{\mathcal{P}}$ is equal to 1 for an even permutation and -1 for an odd permutation. Remember that an even (odd) permutation is a permutation obtainable with an even (odd) number of two-elements swaps. Formally, we can also express $\mathcal{Z}$ in terms of Feynman path integrals as follows,

$$\mathcal{Z}_N = \frac{1}{N!} \sum_{\mathcal{P}} \xi^{\mathcal{P}} \int_{\{\mathbf{r}_i(0)=\mathbf{r}_{\mathcal{P}i}(\hbar\beta)\}} \mathcal{D}[\{\mathbf{r}_i\}]\, e^{-S_E[\{\mathbf{r}_i\}]/\hbar} \tag{2.207}$$

where

$$S_E[\{\mathbf{r}_i\}] = \int_0^{\beta\hbar} d\tau \left[\sum_{i=1}^{N} \left(\frac{1}{2} m\dot{\mathbf{r}}_i^2 + U(\mathbf{r}_i) \right) + \frac{1}{2} \sum_{\substack{i,j=1 \\ i \neq j}}^{N} V(\mathbf{r}_i, \mathbf{r}_j) \right] \tag{2.208}$$

is the classical Euclidean action. For instance, with $N = 2$, we have

$$\mathcal{Z}_2 = \frac{1}{2}\left(\int_{\substack{\mathbf{r}_1(0)=\mathbf{r}_1(\hbar\beta)\\ \mathbf{r}_2(0)=\mathbf{r}_2(\hbar\beta)}} \mathcal{D}[\mathbf{r}_1,\mathbf{r}_2]\, e^{-S_E[\mathbf{r}_1,\mathbf{r}_2]/\hbar} + \int_{\substack{\mathbf{r}_1(0)=\mathbf{r}_2(\hbar\beta)\\ \mathbf{r}_2(0)=\mathbf{r}_1(\hbar\beta)}} \mathcal{D}[\mathbf{r}_1,\mathbf{r}_2]\, e^{-S_E[\mathbf{r}_1,\mathbf{r}_2]/\hbar}\right) \tag{2.209}$$

in the case of two bosons and

$$\mathcal{Z}_2 = \frac{1}{2}\left(\int_{\substack{\mathbf{r}_1(0)=\mathbf{r}_1(\hbar\beta)\\ \mathbf{r}_2(0)=\mathbf{r}_2(\hbar\beta)}} \mathcal{D}[\mathbf{r}_1,\mathbf{r}_2]\, e^{-S_E[\mathbf{r}_1,\mathbf{r}_2]/\hbar} - \int_{\substack{\mathbf{r}_1(0)=\mathbf{r}_2(\hbar\beta)\\ \mathbf{r}_2(0)=\mathbf{r}_1(\hbar\beta)}} \mathcal{D}[\mathbf{r}_1,\mathbf{r}_2]\, e^{-S_E[\mathbf{r}_1,\mathbf{r}_2]/\hbar}\right) \tag{2.210}$$

in the case of two fermions. For a large N it is not an easy task to calculate the many-body partition function (2.206) or (2.207) due to the presence of $N!$ permutations. In the next chapter we will face the same many-body problem in a simpler way: working in the grand canonical ensemble and performing the functional integration over bosonic and fermionic fields.

2.7 Quantum Tunneling

The problem of quantum tunneling can be investigated by using the path integral approach, which is quite useful to determine the tunneling probability of a particle in a classically forbidden region. In Fig. 2.3 we plot the potential energy $U(x)$ for a particle with classical energy E_c. In the classically forbidden region, where $a < x < b$, one has $E_c < U(x)$. We have seen that the saddle point approximation of the Feynman propagator gives

$$K(x_F, t_F|x_I, t_I) = e^{\frac{i}{\hbar}S[x_c]}\,, \tag{2.211}$$

where $x_c(t)$ is the classical trajectory which minimizes the classical action functional with the boundary conditions $x_c(t_I) = x_I$ and $x_c(t_F) = x_F$. Let us consider a classical particle of mass m and coordinate $x(t)$ in the presence of an external potential $U(x)$. We assume that $x = a$ and $x = b$ are two points of the problem for which $U(a) = U(b) = E_c$. In this classically forbidden region, where $a \leq x \leq b$, it is convenient to write

$$p_c = \sqrt{2m(E_c - U(x_c))} = i\sqrt{2m(U(x_c) - E_c)} \,. \tag{2.212}$$

Using Eq. (2.102), the propagator (2.211) in the forbidden region $[a, b]$ gives

$$K(b, t_F|a, t_I) = e^{-(1/\hbar)\int_a^b \sqrt{2m(U(x)-E_c)}\,dx}\, e^{\frac{i}{\hbar}E_c(t_F-t_I)} \,. \tag{2.213}$$

Thus, the probability of tunneling is

$$P_T = |K(b, t_F|a, t_I)|^2 = e^{-(2/\hbar)\int_a^b \sqrt{2m(U(x)-E_c)}\,dx} \,. \tag{2.214}$$

Quite remarkably the same result can be obtained from the stationary Schrödinger equation by adopting the semiclassical Wentzel-Kramers-Brillouin (WKB) approximation. In the next subsection we will discuss the basic properties of the WKB method with some applications.

2.7.1 WKB Method

The WKB method is a semiclassical procedure proposed by Gregor Wentzel, Hendrik Anthony Kramers, and Leon Brillouin in 1926. Given the one-dimensional stationary Schrödinger equation

$$\left(-\frac{\hbar^2}{2m}\frac{d^2}{dx^2} + U(x) \right) \phi(x) = E\, \phi(x) \,, \tag{2.215}$$

the wave function $\phi(x)$ can be written as

$$\phi(x) = A\, e^{\frac{i}{\hbar}\sigma(x)} \,, \tag{2.216}$$

where A is a normalization constant $\sigma(x)$ is a complex function with the units of an action (energy×time). Being a complex number, $\sigma(x)$ gives a contribution both to the modulus and the phase

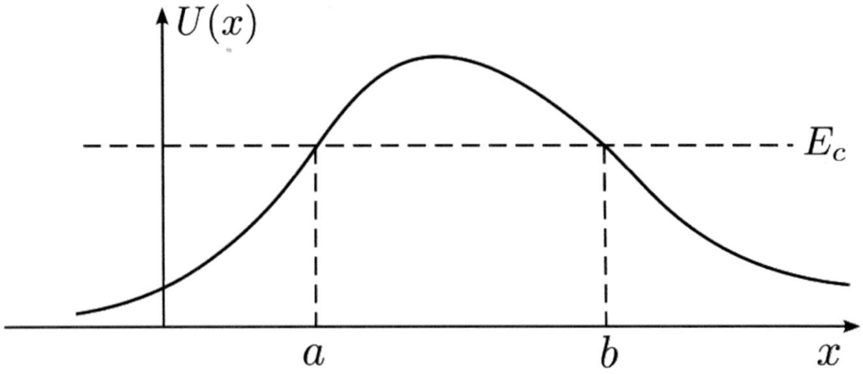

Figure 2.3
Potential barrier $U(x)$ for a particle with classical energy E_c. The region $a \leq x \leq a$ is classically forbidden, but in quantum mechanics the particle can tunnel across the classically forbidden region.

of $\phi(x)$. Inserting Eq. (2.216) into Eq. (2.215), we find a differential equation for $\sigma(x)$, namely

$$(\sigma'(x))^2 + \frac{\hbar}{i}\sigma''(x) = 2m\,(E - U(x)) \ . \tag{2.217}$$

The WKB method takes into account, in a perturbative way, the fact that usually the reduced Planck constant $\hbar$ is much smaller than the modulus $|\sigma(x)|$ of the complex action. Assuming that the term $\hbar\sigma''(x)$ is negligible in Eq. (2.217), we obtain

$$\sigma'(x) = p(x) \tag{2.218}$$

where

$$p(x) = \sqrt{2m\,(E - U(x))} \tag{2.219}$$

is the classical linear momentum. Thus, we have found a remarkable approximate formula which connects the action $\sigma(x)$ to the linear momentum $p(x)$. Eq. (2.218) is the zero-order approximation of the WKB expansion and implies that

$$\sigma(x) = \int_a^x p(x')\,dx' = \int_a^x \sqrt{2m\,(E - U(x'))}\,dx' \tag{2.220}$$

with the initial condition $\sigma(a) = 0$. Thus, in the zero-order WKB approximation the wave function reads

$$\phi(x) = A\, e^{\frac{i}{\hbar}\int_a^x \sqrt{2m(E-U(x'))}\, dx'} \,. \tag{2.221}$$

This expression is very useful to derive the tunneling probability of a particle in a classically forbidden region. If $E \leq U(x)$ for $x \in [a, b]$ with a and b the two turning points, we use Eq. (2.212), and Eq. (2.221) becomes

$$\phi(b) = A\, e^{-\frac{1}{\hbar}\int_a^b \sqrt{2m(U(x)-E)}\, dx} \tag{2.222}$$

in this classically forbidden region. It follows that the probability $P_T = |\phi(b)|^2/|\phi(a)|^2$ of quantum tunneling is given exactly by Eq. (2.214) with the identification $E = E_c$.

The WKB method is extremely powerful, in particular for one-dimensional problems. For instance, it is often used to obtain quite reliable approximate formulas for the quantized energy levels. In general, the single-valuedness of the wave function $\phi(x)$ given by Eq. (2.216) implies that its phase $\sigma(x)/\hbar$ satisfies the equation

$$\frac{1}{\hbar}\oint d\sigma = 2\pi\, n \,, \tag{2.223}$$

where n is a natural quantum number. Within our zero-order approximation, we have

$$\frac{1}{\hbar}\oint d\sigma = \frac{1}{\hbar}\oint \sigma'(x)\; dx = \frac{1}{\hbar}\oint p(x)\; dx \,. \tag{2.224}$$

Thus, the zero order of the WKB expansion with respect to $\hbar$ gives

$$\oint p(x)\, dx = 2\int_a^b \sqrt{2m(E - U(x))}\, dx = h\, n \,, \tag{2.225}$$

that is nothing else than the Bohr-Sommerfeld-Wilson quantization rule of old quantum mechanics, with $h = 2\pi\hbar$ and assuming two turning points a and b such that $U(a) = U(b) = E$. Considering instead the full WKB expansion of the phase $\sigma(x)$, that is

$$\sigma(x) = \sum_{k=0}^{+\infty} (\frac{\hbar}{i})^k\, \sigma_k(x) \,, \tag{2.226}$$

after substituting (2.226) into (2.217) and comparing like powers of $\hbar$, one finds the recursion relation ($n > 0$)

$$\sigma''_{n-1}(x) + \sum_{k=0}^{n} \sigma'_k(x)\, \sigma'_{n-k}(x) = 0 \tag{2.227}$$

with initial condition

$$\sigma'_0(x) = \sqrt{2m\,(E - U(x))} = p(x)\ . \tag{2.228}$$

For $n = 1$ one gets

$$\sigma''_0 + 2\sigma'_0\ \sigma'_1 = 0, \tag{2.229}$$

i.e.

$$\sigma'_1(x) = -\frac{1}{2}\frac{\sigma''_0(x)}{\sigma'_0(x)} = -\frac{1}{2}\frac{d}{dx}\left(\ln \sigma'_0(x)\right) = -\frac{1}{2}\frac{d}{dx}\left(\ln p(x)\right)\ . \tag{2.230}$$

The quantization condition, obtained by requiring the single-valuedness of the wave function, gives

$$\oint d\sigma = \sum_{k=0}^{+\infty} (\frac{\hbar}{i})^k \oint d\sigma_k = 2\pi\hbar\, n\ . \tag{2.231}$$

The zero-order contribution is exactly given by Eq. (2.225). Instead, the first odd term relates to Eq. (2.230), from which

$$d\sigma_1 = -\frac{1}{2} d\left(\ln p(x)\right)\ . \tag{2.232}$$

Moreover, setting $p(x) = e^{i\alpha(x)}$ one finds $\ln(p(x)) = i\alpha(x)$ and also

$$\oint d\left(\ln p(x)\right) = i \oint d\alpha(x) = i \int_0^{2\pi} d\alpha = 2\pi i\ . \tag{2.233}$$

In this way, we obtain the so-called Maslov correction

$$(\frac{\hbar}{i}) \oint d\sigma_1 = -(\frac{\hbar}{i})\frac{1}{2} \oint d\left(\ln p(x)\right) = -\pi\hbar = -\frac{h}{2}\ . \tag{2.234}$$

We conclude that, to the first order in the WKB expansion of $\hbar$, one has

$$\oint p(x)\, dx = h\left(n + \frac{1}{2}\right)\ , \tag{2.235}$$

namely

$$2\int_a^b \sqrt{2m(E-U(x))}\,dx = h\left(n+\frac{1}{2}\right)\,, \tag{2.236}$$

in the case of two turning points.

Another relevant application of the WKB method is the analytical detemination of the energy splitting ΔE of two adjacent energy levels in the problem of a double-well potential. As an example, in Fig. 2.4 we plot a double-well potential

$$U(x) = A(x^2 - x_0^2)^2\,. \tag{2.237}$$

At fixed energy E, the classically forbidden region holds for $-a < x < a$ while the classically allowed region is characterized by $-b \le x \le -a$ or $a \le x \le b$, where $\pm a$ and $\pm b$ are the four solutions (turning points) of the equation $E = U(x)$.

We consider the one-dimensional system with a quantum Hamiltonian

$$\hat{H} = -\frac{\hbar^2}{2m}\frac{d^2}{dx^2} + U(x)\,, \tag{2.238}$$

where $U(-x) = U(x)$ is the symmetric double-well potential. Let ϕ_1 and ϕ_2 be two exact eigenfunctions of the stationary Schrödinger equation, i.e.

$$\hat{H}\psi_1 = E_1\phi_1 \qquad \text{and} \qquad \hat{H}\psi_2 = E_2\phi_2\,, \tag{2.239}$$

such that $\phi_1(-x) = \phi_1(x)$, $\phi_2(-x) = -\phi_2(x)$, and $E_1 \simeq E_2$. To calculate the splitting

$$\Delta E = E_2 - E_1 \tag{2.240}$$

we multiply the first equation by $\phi_2(x)$ and the second by $\phi_1(x)$ and then we subtract the two resulting equations. By integrating we find

$$\Delta E = \frac{\hbar^2}{2m}\frac{\phi_1(0)\phi_2'(0) - \phi_1'(0)\phi_2(0)}{\int_0^{+\infty}\phi_1(x)\phi_2(x)\,dx}\,. \tag{2.241}$$

We write the eigenfunctions $\phi_1(x)$ and $\phi_2(x)$ in terms of the right-localized function

$$\phi_R(x) = \frac{1}{\sqrt{2}}(\phi_1(x) + \phi_2(x))\,. \tag{2.242}$$

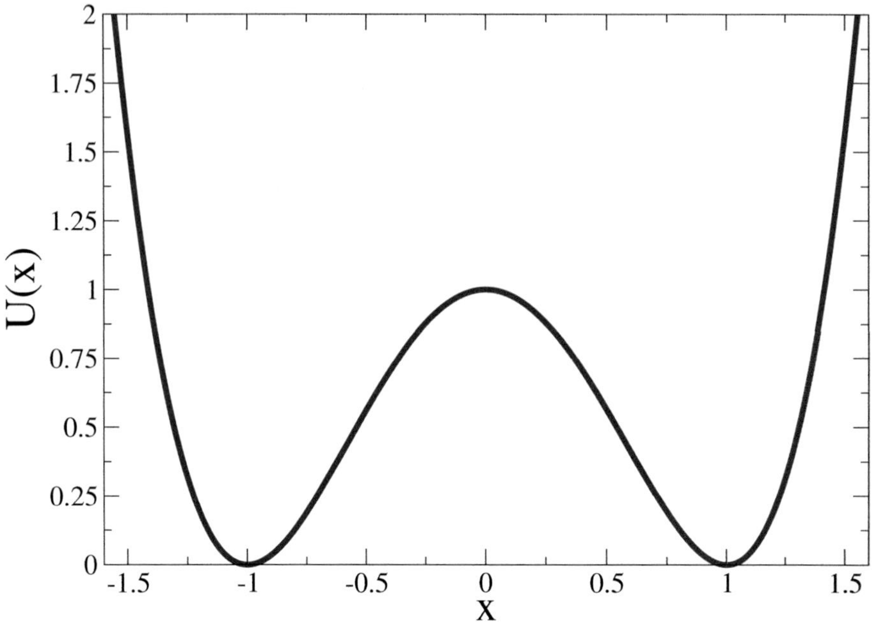

Figure 2.4
Double-well potential $U(x) = A\,(x^2 - x_0^2)^2$ with $A = 1$ and $x_0 = 1$. In this plot the maximum $U_0 = Ax_0^4 = 1$ is located at $x = 0$ and the two minima are at $x = \pm x_0 = \pm 1$. The classically forbidden region is caracterized by $E < U_0$, where E is the total energy of the particle is traveling along the classical path.

It is easy to show that $E_R = \langle \phi_R | \hat{H} | \phi_R \rangle = (E_1 + E_2)/2$. Then, we get

$$\Delta E = \frac{2\hbar^2}{m} \psi_R(0) \psi'_R(0) \,, \tag{2.243}$$

which is an almost exact starting formula to calculate the energy splitting. By using the WKB method, at the first order, we have

$$\phi_R(x) = \frac{C}{\sqrt{2m(U(x) - E)}} e^{-\frac{1}{\hbar} \int_x^a \sqrt{2m(U(x') - E)}\, dx'} \tag{2.244}$$

in the classically forbidden region close to $x = 0$. The normalization constant is obtained by imposing

$$1 \simeq \int_0^{+\infty} |\psi_R(x)|^2 \simeq 2C^2 \int_a^b \sqrt{2m(E - U(x))}\, dx \,. \qquad (2.245)$$

Here the main contribution to the integral is due to the wavefunction $\phi_R(x)$ in the classically allowed region $[a, b]$, and the factor 2 comes from the combination of two plane waves with opposite directions in this region. Combining all the results, we finally obtain

$$\Delta E = \hbar\omega_{\rm a}\, e^{-S_{\rm f}/\hbar} \,, \qquad (2.246)$$

where

$$\omega_{\rm a} = \frac{1}{m} \frac{1}{\int_a^b \sqrt{2m(E - U(x))}\, dx} \qquad (2.247)$$

is the angular frequency of oscillation in the classically allowed region and

$$S_{\rm f} = \int_{-a}^{a} \sqrt{2m(U(x) - E)}\, dx \qquad (2.248)$$

is the action in the classically forbidden region. Notice that

$$\Gamma_T = \frac{\Delta E}{2\pi\hbar} = \frac{\omega_{\rm a}}{2\pi}\, e^{-S_{\rm f}/\hbar} \qquad (2.249)$$

can be interpreted as the tunneling probability rate from the right well to the left well.

2.7.2 Imaginary-Time Approach

It is important to stress that in the scientific literature, within a path-integral formulation, quantum tunneling is usually discussed working directly with imaginary time, setting

$$t_F - t_I = -i(\tau_F - \tau_I) \qquad (2.250)$$

from which Eq. (2.212) follows. This approach formally corresponds to invert the sign of the potential: $U(x) \to -U(x)$. In this case the two turning points of the classically forbidden region are obtained setting $E_c = U(x)$ and we work with $x_I = a$ and $x_F = b$.

Clearly the classically forbidden region corresponds to $a \leq x \leq b$. Moreover, we have $x_c(\tau_I) = a$, $x_c(\tau_F) = b$, and $\dot{x}_c(\tau_I) = \dot{x}_c(\tau_F) = 0$. Taking into account all these assumptions and also Eq. (2.102), the propagator (2.211) in the forbidden region $[a, b]$ gives

$$K(b, \tau_F|a, \tau_I) = e^{-(1/\hbar)\int_a^b \sqrt{2m(U(x)-E_c)}\,dx}\, e^{-E_c(\tau_F-\tau_I)/\hbar} \tag{2.251}$$

and the probability of tunneling reads

$$P_T = |K(b, \tau_F|a, \tau_I)|^2 = e^{-2E_c(\tau_F-\tau_I)/\hbar}\, e^{-(2/\hbar)\int_a^b \sqrt{2m(U(x)-E_c)}\,dx} . \tag{2.252}$$

Notice that under the short-time assumption $E_c(\tau_F - \tau_I) \ll \hbar$ Eq. (2.252) becomes

$$P_T = |K(b, \tau_F|a, \tau_I)|^2 = e^{-(2/\hbar)\int_a^b \sqrt{2m(U(x)-E_c)}\,dx} , \tag{2.253}$$

that is sometimes called a single-instanton result, and it is precisely Eq. (2.214). In this imaginary time approach, the classical energy E_c depends on the position a at time τ_I and on the final position b at time τ_F of the two turning points. In fact, from Eq. (2.95) we have

$$\sqrt{\frac{2}{m}(U(x_c) - E_c)} = \frac{dx_c}{d\tau} \tag{2.254}$$

from which

$$\int_{\tau_I}^{\tau_F} d\tau = \int_a^b \frac{dx_c}{\sqrt{\frac{2}{m}(U(x_c) - E_c)}} \tag{2.255}$$

and then

$$(\tau_F - \tau_I) = \sqrt{\frac{m}{2}} \int_a^b \frac{dx}{\sqrt{U(x) - E_c}} \tag{2.256}$$

that is an implicit formula for the classical energy E_c. Actually, the classical solution is periodic, due to the infinite bounces between the two turning points, and the single-instanton (or single-kink) corresponds to only one period of the imaginary-time periodic solution. An improvement of Eq. (2.253) is obtained by including all the bounces (termed multi-instanton configuration or instanton gas) of the classical solution. Eq. (2.253), or more generally

Eq. (2.252), could be further improved, taking into account the Gaussian fluctuations.

A simple analytical solution of our double-well problem is obtained in the very special case that the two turning points $x = -a$ and $x = +a$ coincide with the two minima $x = -x_0$ and $x = +x_0$ of the potential (2.237). In this case the classical energy E_c has its minimal value $E_c = 0$, and the time interval $\tau_F - \tau_I$ of Eq. (2.256) goes to infinity. The classical trajectory is again derived from Eq. (2.254) integrating in the time interval $[0, \tau]$, namely

$$\tau = \int_0^{\tau} d\tau' = \int_0^{x_c(\tau)} \frac{dx}{\sqrt{\frac{2A}{m}(x_0^2 - x^2)^2}} = \sqrt{\frac{m}{2A}} \text{arctanh}\left(\frac{x_c(\tau)}{x_0}\right) . \tag{2.257}$$

In this way we obtain

$$x_c(\tau) = x_0 \tanh\left(\sqrt{\frac{2A}{m}}\,\tau\right) . \tag{2.258}$$

This very special classical solution is also called an instanton (or kink) because the time interval $\sqrt{m/(2A)}$ in which $x_{cl}(\tau)$ remains around $x = 0$ is extremely small compared with the (infinite) long time the solution $x_c(\tau)$ stays at $x = -x_0$ for $\tau < 0$ and at $x = +x_0$ for $\tau > 0$. The anti-instanton (or anti-kink) configuration is obtained by inverting the sign of τ. The Euclidean action functional calculated with $x_c(\tau)$ of Eq. (2.258) is nothing else than

$$\begin{aligned} S_E[x_c] &= \int_{-x_0}^{+x_0} dx\, \sqrt{2m(A(x_0^2 - x^2)^2)} \\ &= \sqrt{2mA} \int_{-x_0}^{+x_0} dx\, (x_0^2 - x^2) \\ &= \frac{4}{3} x_0^3 \sqrt{2mA} . \end{aligned} \tag{2.259}$$

Having the action $S[x_c]$, we get

$$P_T = |K(+x_0, +\infty| - x_0, -\infty)|^2 = e^{-4x_0^3\sqrt{2mA}/(3\hbar)} , \tag{2.260}$$

that is once again, the semiclassical tunneling probability formula, this time applied to the double-well problem. For the sake of completeness, we emphasize that the instantonic solution (2.258) is not the most general of Eq. (2.254) with boundary conditions $x(-\infty) = -x_0$ and $x(+\infty) = x_0$. A more general solution is

$$x(\tau) = x_0 \tanh\left(\sqrt{\frac{2A}{m}}\,(\tau - \tau_0) \right) , \tag{2.261}$$

where $\tau_0 \in \mathbb{R}$ is an arbitrary parameter fixing the characteristic time of jumping from $-x_0$ to x_0, i.e., the center of the instanton solution. Actually, the most general solution $x_n(\tau)$ can have n different jumps at n times τ_n, i.e., $n-1$ bounces back and forth, before reaching the final state. Summing over all the possible $x_n(\tau)$, that is, summing over the possible instantons, one gets an improved determination of the tunneling probability.

2.7.3 Escape Tunneling

The imaginary time approach can be used to calculate the semiclassical probability of escape tunneling in the case of the cubic potential

$$U(x) = \frac{m\omega^2}{2} x^2 (1 - \alpha\, x) \tag{2.262}$$

with $\alpha \geq 0$. For $\alpha = 0$ one recovers the familiar harmonic potential, and at $x = 0$ there is local and absolute minimum of the system. Instead for $\alpha > 0$ at $x = 0$ there a local minimum, at $x = 2/(3\alpha)$ there is a local maximum, but the absolute minimum is at $x = +\infty$. See Fig. 2.5.

For $\alpha > 0$ it is quite simple to analyze the quantum tunneling from $x = 0$ to $x = 1/\alpha$ that are the two turning points when the classical energy E_c is zero. As before, the classical trajectory is obtained from

$$\sqrt{2}\,\omega\,\tau = \int_0^{x_c(\tau)} \frac{dx}{\sqrt{x^2(1 - \alpha\, x)}} = 2\,\text{arctanh}[\sqrt{1 - \alpha\, x_c(\tau)}] \, . \tag{2.263}$$

In this way we find

$$x_c(\tau) = \frac{1}{\alpha} \left[1 - \tanh^2(\frac{\omega\tau}{\sqrt{2}}) \right] , \tag{2.264}$$

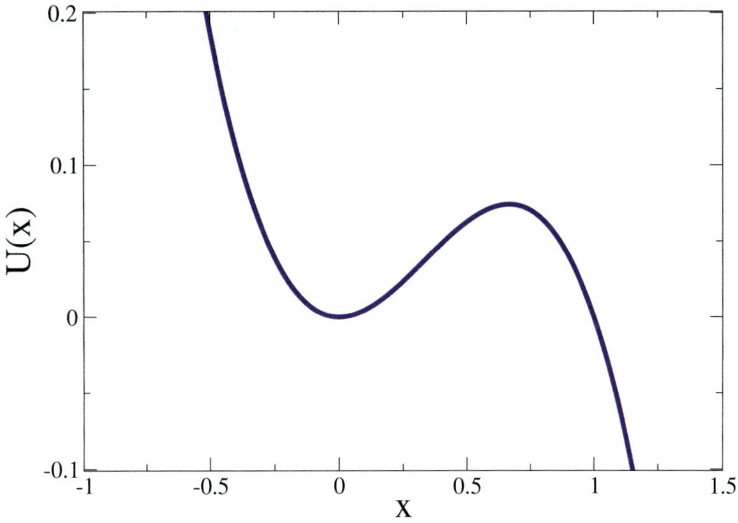

Figure 2.5
Cubic potential $U(x) = m\omega^2 x^2(1-\alpha x)/2$ setting $m\omega^2 = 1$ and $\alpha = 1$.

where, as expected, $x_c(0) = 0$ and $x_c(+\infty) = 1/\alpha$. The Euclidean action calculated on this classical trajectory, where $x_I = 0$ and $x_F = 1/\alpha$, reads

$$S_E[x_c] = \int_0^{\frac{1}{\alpha}} dx\, \sqrt{m^2\omega^2 x^2(1-\alpha\, x)} = m\,\omega \int_0^{\frac{1}{\alpha}} dx\, x\sqrt{1-\alpha\, x}$$
$$= \frac{2m\omega}{15\alpha^2}\,. \tag{2.265}$$

It follows that the tunneling probability is given by

$$P_T = |K(\frac{1}{\alpha}, +\infty|0,0)|^2 = e^{-4m\omega/(15\hbar\alpha^2)} \tag{2.266}$$

at the lowest order of the semiclassical approximation, i.e., the saddle-point approximation with a single instanton and without Gaussian fluctuations. As suggested for the first time (1928) by George Gamow, in the context of nuclear decay of α particles, the tunneling probability rate Γ_T can be written as

$$\Gamma_T = \frac{\omega_a}{2\pi}\, P_T\,, \tag{2.267}$$

where ω_a is the angular frequency of oscillation of the particle in the classically allowed region. This is the same kind of expression, we have found for the double-well problem.

2.8 Classical Stochastic Processes

Remarkable similarities exist between quantum mechanics and the theory of classical stochastic processes. In particular, the quantum probability amplitudes satisfy formal rules almost identical to those that apply to the probabilities of a classical Markov process. Considering the time-dependent Schrödinger equation (2.8) of a free particle, i.e., $U(x) = 0$, and performing the Wick rotation $t = -i\tau$ we get

$$\frac{\partial}{\partial\tau}\psi(x,\tau) = \frac{\hbar^2}{2m}\frac{\partial^2}{\partial x^2}\psi(x,\tau) \; . \tag{2.268}$$

With the identifications $\psi(x,\tau) = P(x,\tau)$ and $D = \hbar^2/(2m)$, this imaginary-time equation is formally equivalent to the diffusion equation (Fick law)

$$\frac{\partial}{\partial\tau}P(x,\tau) = D\frac{\partial^2}{\partial x^2}P(x,\tau) \; , \tag{2.269}$$

where $P(x,\tau)$ is the classical probability of finding a particle in the position x at time τ, with D the so-called diffusion constant. Eq. (2.269) is a simple example of the Fokker-Planck equation describing a Markov stochastic process, where the probability of each event (in a sequence of possible events) depends only on the state attained in the previous event. Imposing the initial condition

$$P(x,\tau_I) = \delta(x - x_I) \tag{2.270}$$

and also the normalization

$$\int_{-\infty}^{+\infty} P(x,\tau)\, dx = 1 \tag{2.271}$$

for any τ, Eq. (2.269) has the solution

$$P(x,\tau) = \frac{1}{\sqrt{4\pi D(\tau - \tau_I)}}\, e^{-\frac{1}{4D}\frac{(x-x_I)^2}{(\tau-\tau_I)}} . \qquad (2.272)$$

The meaning of this solution is that, due to the stochastic diffusion (Brownian motion), the probability of finding a particle spreads around the initial position x_I with a time-dependent Gaussian distribution. In particular, one finds that

$$\langle x^2 \rangle = \int_{-\infty}^{+\infty} x^2\, P(x,\tau) = x_I^2 + 2D(\tau - \tau_I) . \qquad (2.273)$$

Needless to say, the solution (2.272) is nothing else than the propagator (2.48) of a free quantum particle with imaginary time and $D = \hbar^2/(2m)$. As discussed at the beginning of this chapter, Wiener was the first to develop a path integral approach (that is, an imaginary-time path integral approach seen from the perspective of quantum mechanics) with the aim of investigating stochastic processes. It was Mark Kac in 1949 to recognize the similarities between the Feynman path integral (1948) and the Wiener stochastic integral (1921). Going further back in time, explaining the Brownian (random) motion of a particle suspended in a fluid, Einstein found (1905) the remarkable formula $D = k_B T/\gamma$ for the diffusion constant of the particle that is assuumed to be in contact with a thermal bath at temperature T and under the effect of both a viscous force $-\gamma\dot{x}$ and a stochastic force.

2.8.1 Langevin Equation and Path Integral

One of the most important equations in the field of stochastic processes is the Langevin equation. Let us analyze some aspects and connections with the path integral formulation. We consider a particle of mass m and coordinate $x(t)$ at (real) time t under the action of a deterministic potential $U(x)$ and in contact with a thermal bath of temperature T which produces a dissipative force $-\gamma\dot{x}(t)$ with damping coefficient γ and a Gaussian stochastic force $\xi(t)$. The coordinate $x(t)$ of the particle satisfies the stochastic equation of motion

$$m\,\ddot{x}(t) + \gamma\,\dot{x}(t) + U'(x(t)) = \xi(t) , \qquad (2.274)$$

where $\dot{x} = dx/dt$ is the derivative with respect to time t and $U' = dU/dx$ is the derivative with respect to x. Equation (2.274) is called classical Langevin equation provided that the probability functional of the stochastic variable $\xi(t)$ with $t \in [t_I, t_F]$ is given by

$$P[\xi] = \exp\left(-\frac{1}{4\gamma k_B T}\int_{t_I}^{t_F} \xi(t)^2\, dt\right) \tag{2.275}$$

with k_{B} the Boltzmann constant. The presence of the damping coefficient γ in both Eqs. (2.274) and (2.275) is needed to recover at equilibrium the results of classical statistical mechanics (fluctuation-dissipation theorem). In full generality, given a quantity A that is a functional of $\xi(t)$, i.e., $A = A[\xi]$, its stochastic average is defined as

$$\langle A\rangle = \int D[\xi]\, A[\xi]\, P[\xi]\,. \tag{2.276}$$

Taking into account Eq. (2.275) and the relationship between $\xi(t)$ and $x(t)$, that is Eq. (2.274), i.e., substituting $\xi(t)$ with $x(t)$ into Eq. (2.275) by using Eq. (2.274), we obtain

$$P[x] = \exp\left(-\frac{F[x]}{4\gamma k_B T}\right), \tag{2.277}$$

where

$$\mathcal{F}[x] = \int_{t_I}^{t_F} \left[m\,\ddot{x}(t) + \gamma\,\dot{x}(t) + U'(x(t))\right]^2 dt \tag{2.278}$$

is the so-called Onsager-Machlup functional, and the generic path $x(t)$ must have the end points $x(t_I) = x_I$ and $x(t_F) = x_F$. Notice that the Jacobian of the change of variables $\xi(t) \to x(t)$ is absorbed in the measure $D[x]$.

In the overdamped regime, where γ is large and $m\ddot{x}$ is negligible, the Langevin equation (2.274) becomes

$$\gamma\,\dot{x}(t) + U'(x(t)) = \xi(t) \tag{2.279}$$

and the Onsager-Machlup functional reads

$$\begin{aligned}\mathcal{F}[x] &= \int_{t_I}^{t_I} \left[\gamma\,\dot{x}(t) + U'(x(t))\right]^2\,dt \\ &= \int_{t_I}^{t_F} \left[\gamma^2\,\dot{x}(t)^2 + U'(x(t))^2\right]\,dt + 2\gamma\Delta U \end{aligned} \qquad (2.280)$$

with

$$\Delta U = U(x_F) - U(x_I) \qquad (2.281)$$

and taking into account that

$$\dot{x}\,U'\,dt = \frac{dx}{dt}\,\frac{dU}{dx}\,dt = dU\;. \qquad (2.282)$$

This last formula, that is valid within the familiar deterministic calculus of Newton-Leibnitz, also holds in the framework of the stochastic calculus *à la* Stratonovich.

2.8.2 Saddle Point and Arrhenius Law

It is important to observe that from the probability functional (2.277) one finds the conditional probability $P(x_F, t_I|x_I, t_I)$ of finding the particle in the position x_F at time t_F with the constraint of being in the position x_I at the initial time t_I, i.e.

$$P(x_F, t_I|x_I, t_I) = \int_{x(t_I)=x_I}^{x(t_F)=x_F} \mathcal{D}[x]\,\exp\left(-\frac{F[x]}{4\gamma k_B T}\right) \qquad (2.283)$$

In the overdamped regime and for $\gamma k_B T$ sufficiently large, one can adopt for (2.277) the saddle point approximation

$$P(x_F, t_F|x_I, t_I) = e^{-\mathcal{F}[x_c]/(4\gamma k_B T)} \qquad (2.284)$$

with $x_c(t)$ is the trajectory which minimizes the Onsager-Machlup functional. This trajectory $x_c(t)$ satisfies the differential equation

$$2\gamma^2\ddot{x}_c(t) + V'(x_c(t)) = 0\;, \qquad (2.285)$$

where

$$V(x) = -U'(x)^2 \qquad (2.286)$$

plays the role of an effective potential, while $2\gamma^2$ is a sort of effective mass. Eq. (2.285) has the following constant of motion

$$K = \gamma^2 \dot{x}_c(t)^2 + V(x_c(t)) = \gamma^2 \dot{x}_c(t)^2 - U'(x_c(t))^2 \qquad (2.287)$$

during the time evolution of $x_c(t)$. Taking into account that

$$dt = \frac{dx}{\dot{x}} = \frac{\gamma\, dx}{\sqrt{K + U'(x)^2}} \,, \qquad (2.288)$$

on the classical trajectory $x_c(t)$ the Onsager-Machlup functional reads

$$\mathcal{F}[x_c] = \int_{x_I}^{x_F} 2\gamma \sqrt{K + U'(x)^2}\, dx - K(t_F - t_I) + 2\gamma\, \Delta U \,. \quad (2.289)$$

We now consider the problem of classical tunneling activated by thermal effects, where $\Delta U > 0$ is the energy barrier of the potential $U(x)$ with x_I the position of the local minimum and x_F the position local maximum. We set $U(x_I) = 0$ and $K = 0$. It follows that the Onsager-Machlup is strongly simplified:

$$\mathcal{F}[x_c] = 2\gamma \int_{x_I}^{x_F} \sqrt{U'(x)^2}\, dx + 2\gamma\, \Delta U = 4\gamma \Delta U \,. \qquad (2.290)$$

Finally, inserting into Eq. (2.284) we obtain

$$P_T = e^{-\Delta U/(k_B T)} \,, \qquad (2.291)$$

that is the Arrhenius law of the classical tunneling probability induced by thermal excitations. Notice that the viscosity γ drops out within this saddle-point scheme. At fixed energy barrier ΔU by increasing the temperature the probability P_T of tunneling is enhanced, a quite intuitive but meaningful result.

2.9 Solved Problems

Problem 2.1

Prove the formula

$$\int_{-\infty}^{+\infty} dx\, e^{-\frac{a}{2}x^2 + bx} = \sqrt{\frac{2\pi}{a}}\, e^{\frac{b^2}{2a}} \,,$$

knowing that

$$\int_{-\infty}^{+\infty} dx\, e^{-x^2} = \sqrt{\pi}\ .$$

Solution

By completing the square in the exponent of the exponential, we have

$$\int_{-\infty}^{+\infty} dx\, e^{-\frac{a}{2}x^2+bx} = e^{\frac{b^2}{2a}} \int_{-\infty}^{+\infty} dx\, e^{-(\sqrt{\frac{a}{2}}x-\frac{b}{2}\sqrt{\frac{2}{a}})^2}\ .$$

Changing variable

$$y = \sqrt{\frac{a}{2}}x - \frac{b}{2}\sqrt{\frac{2}{a}}$$

we obtain

$$\int_{-\infty}^{+\infty} dx\, e^{-(\sqrt{\frac{a}{2}}x-\frac{b}{2}\sqrt{\frac{2}{a}})^2} = \sqrt{\frac{2}{a}} \int_{-\infty}^{+\infty} dy\, e^{-y^2} = \sqrt{\frac{2\pi}{a}}\ .$$

Finally, we get

$$\int_{-\infty}^{+\infty} dx\, e^{-\frac{a}{2}x^2+bx} = e^{\frac{b^2}{2a}} \sqrt{\frac{2\pi}{a}}$$

which is the desired result.

Problem 2.2

In the high-temperature limit, calculate the partition function of a particle of mass m confined by the quartic potential

$$U(x) = A\,x^4$$

with A a positive constant and $x(t)$ the coordinate of the particle.

Solution

We have seen that for a sufficiently high temperature T the partition function reads

$$\mathcal{Z}_1 = \sqrt{\frac{mk_BT}{2\pi\hbar^2}} \int_{-\infty}^{+\infty} dx_I\, e^{-\beta U(x_I)}\ .$$

In our case

$$\mathcal{Z}_1 = \sqrt{\frac{mk_BT}{2\pi\hbar^2}} \int_{-\infty}^{+\infty} dx_I \, e^{-\beta A x_I^4} \, .$$

The integral can be analytically evaluated, giving

$$\int_{-\infty}^{+\infty} dx_I \, e^{-\beta A x_I^4} = \frac{2\,\Gamma(5/4)}{\beta\, A}$$

with $\Gamma(x)$ the Euler gamma function. Thus, the partition function is given by

$$\mathcal{Z}_1 = \sqrt{\frac{2m}{\pi\hbar^2}\frac{\Gamma(5/4)}{A}}(k_BT)^{3/2} \, .$$

Problem 2.3

Consider a particle of mass m colliding with energy E on a square barrier of amplitude U_0 and width d. Determine the tunneling probability by using the saddle-point approximation.

Solution

In the saddle point approximation, the tunneling probability is given by

$$P_T = e^{-(2/\hbar)\int_{x_I}^{x_F} \sqrt{2m(U(x)-E)}\, dx} \, .$$

In our problem the potential energy is

$$U(x) = \begin{cases} U_0 & \text{for } -\frac{d}{2} \leq x \leq \frac{d}{2} \\ 0 & \text{elsewhere} \end{cases}$$

and we choose $x_I = -d/2$ and $x_F = d/2$. It then follows immediately

$$P_T = e^{-(2/\hbar)d\sqrt{2m(U_0-E)}} \, .$$

Problem 2.4

By using the WKB method, calculate the energy spectrum of the harmonic oscillator.

Solution

In the case of the harmonic oscillator, whose potential energy is

$$U(x) = \frac{1}{2} m\omega^2 x^2 \; ,$$

we have

$$\begin{aligned} \oint p(x)\, dx &= 2\int_{-a}^{a} \sqrt{2m(E - \frac{1}{2}m\omega^2 x^2)}\, dx \\ &= 2\sqrt{2mE}\; a\, \frac{\pi}{2} = \pi\sqrt{2mE}\sqrt{\frac{2E}{m\omega^2}} = 2\pi\frac{E}{\omega} \end{aligned}$$

with $-a$ and $a = \sqrt{2E/(m\omega^2)}$ are the two turning points for which $E = U(\pm a)$. From Eq. (2.235), we then obtain

$$2\pi\frac{E}{\omega} = h\left(n + \frac{1}{2}\right) \; ,$$

namely

$$E = \hbar\omega\left(n + \frac{1}{2}\right) \; ,$$

which is the exact quantization formula of the harmonic oscillator, remembering that $\hbar = h/(2\pi)$. Notice that in this specific case, all higher-order WKB corrections are exactly zero.

Problem 2.5

For a Schwarzschild black hole of mass M, which is spherical, not rotating, and electrically neutral, its Bekenstein-Hawking entropy is given by

$$\mathcal{S} = \frac{4\pi k_B}{\hbar}\frac{GM^2}{c} \; ,$$

where G is the Newton gravitational constant, $\hbar$ is the reduced Planck constant, and c is the speed of light in a vacuum. Assuming that $E = Mc^2$ is the internal energy of the black hole, derive the Hawking temperature T of the black hole and also its Helmholtz free energy F.

Solution

Under the assumption that the internal energy E of the system is only the rest energy due to the mass M of the black hole, i.e.

$$E = Mc^2 ,$$

we can rewrite the entropy formula (proposed for the first time in 1972 by Jacob Bekenstein and refined in 1974 by Stephen Hawking) as follows

$$\mathcal{S} = \frac{4\pi k_B G}{\hbar c^5} E^2 .$$

Having the entropy $\mathcal{S}$ as a function of the internal energy, the temperature T can be derived using

$$\frac{1}{T} = \frac{\partial \mathcal{S}}{\partial E} .$$

It gives

$$\frac{1}{T} = \frac{8\pi k_B G}{\hbar c^5} E$$

Equivalently, by using again $E = Mc^2$, we get

$$T = \frac{\hbar c^3}{8\pi k_B G M} ,$$

that is the so-called Hawking temperature. Finally, the Helmholtz free energy, which is defined as

$$F = E - T\mathcal{S} ,$$

can be easily calculated, and it reads

$$F = \frac{c^5 \hbar}{16\pi G k_B T}$$

in terms of the temperature T.

Further Reading

The standard reference of the path integral formulation is
– R.P. Feynman, A.R. Hibbs, and D.F. Styer, Quantum Mechanics and Path Integrals. Emended Edition (Dover, New York, 2010).
where many physically relevant examples are discussed. A pedagogical approach to the path integral can be found in
– M. Ogilvie, A Multidisciplinary Approach to Quantum Field Theory. An introduction (IOP Publishing, Bristol, 2022).
A comprehensive discussion of the Feynman path integral with several applications is given in these books:
– H. Kleinert, Path Integrals In Quantum Mechanics, Statistics, Polymer Physics, And Financial Markets (World Scientific, Singapore, 2009).
– A. Altland and B. Simons, Condensed Matter Field Theory (Cambridge Univ. Press, 2010).
– J. Zinn-Justin, Quantum Field Theory and Critical Pheonomena (Oxford Univ. Press, 2021).

3

Path Integral for Bosons

In this chapter we consider the quantum statistical mechanics of identical nonrelativistic bosons by introducing the formalism of functional integration for the bosonic field. We then derive the equation of state of non-interacting and weakly-interacting massive bosons taking into account beyond-mean-field regularized Gaussian fluctuations. We also analyze various properties of the mean-field time-dependent Gross-Pitaevskii equation. In particular, we show that this equation is equivalent to the zero-temperature equations of superfluid hydrodynamics proposed in 1949 by Landau. Finally, we discuss topological and solitonic solutions of the Gross-Pitaevskii equation.

3.1 Introduction

Among phase transitions, very peculiar are the superfluid transition of liquid helium and the superconducting transition in many materials. These transitions are a macroscopic-scale manifestation of quantum mechanics. Both superfluidity and superconductivity are emerging phenomena, originating from the formation of a Bose-Einstein condensate (or a quasi-condensate) at the microscopic level.

If one considers an ideal gas of atoms at temperature T, based on Maxwell-Boltzmann kinetic theory of gases, the kinetic energy average of an atoms is linked to the temperature T by the formula

$$\frac{1}{2}m\langle v^2\rangle = \frac{3}{2}k_B T \ , \tag{3.1}$$

where k_B is the Boltzmann constant, and the symbol $\langle ... \rangle$ means

DOI: 10.1201/9781003588337-3

thermal average. Then we have that

$$\bar{v} = \langle v^2 \rangle^{1/2} = \sqrt{\frac{3k_BT}{m}} \,. \tag{3.2}$$

According to quantum mechanics, and in particular according to the wave theory of Louis de Broglie, to each particle of mass m and velocity $\bar{v}$ is associated with a wave whose length λ_T is given by

$$\lambda_T = \frac{h}{m\bar{v}} \,, \tag{3.3}$$

where h is Planck's constant. The de Broglie wavelength λ_T can be then write, using (3.2), as

$$\lambda_T = \frac{h}{\sqrt{3mk_BT}} \,. \tag{3.4}$$

This formula shows that the length λ_T increases as it decreases of the temperature T and diverges when the temperature (measured in degrees Kelvin) approaches absolute zero. There is therefore a temperature T^* below which λ_T becomes equal to the distance average d between atoms. If N is the number of atoms and V is the volume of the gas, then the average particle density is $n = N/V$ and the interatomic distance is $d = 1/n^{1/3}$. The temperature T^* can therefore be estimated by putting

$$\frac{\lambda_T}{d} = \lambda n^{1/3} \simeq 1 \,, \tag{3.5}$$

from which, by substituting Eq. (3.4), we obtain

$$T^* \simeq \frac{h^2 n^{2/3}}{3mk_B} \,. \tag{3.6}$$

For $T < T^*$ the quantum waves of different atoms begin to overlap, the atoms become *indistinguishable*, and therefore the kinetic theory of Maxwell-Boltzmann classical is no longer applicable. In this case we talk about *quantum degeneracy*. Usually in gases the density n is so low that the temperature T^* is close to zero absolute. On the other hand, decreasing the temperature at constant pressure, a gas tends to pass into a liquid state and then in solid state.

At atmospheric pressure (1 atm), helium is the only substance that it does not become solid but remains liquid at absolute zero. Despite this, in 1995 Eric Cornell and Carl Wieman at the NIST-JILA laboratory at the University of Colorado and Wolfgang Ketterle at MIT in Boston managed to obtain very dilute atomic vapors (respectively of rubidium and sodium) which remain gaseous up to the temperature of 30 nK (nanokelvin) for several minutes in a meta-stable state. These pioneering experiments paved the way to many subsequent studies of macroscopic quantum properties by using dilute and ultracold gases made of atomic bosons or atomic fermions.

In the first chapter we have shown that quantum statistical mechanics demonstrates that for an ideal gas of identical bosons the distribution of energy E is given by the *Bose-Einstein distribution*

$$f(E) = \frac{1}{e^{(E-\mu)/(k_B T)} - 1} . \tag{3.7}$$

The chemical potential μ is the chemical potential that sets the normalization of the distribution, i.e., the total number N of particles. We have also seen that this distribution is nothing else that distribution derived by the Max Planck to explain the electromagnetic radiation of the black body, with $\mu = 0$. In 1925 Einstein generalized the Planck distribution to the case of particles with mass, based on the work of 1924 by Bose.

In the next sections of this chapter, we will discuss in detail the physics of Bose-Einstein condensation by using the formalism of the path integral. To do this we will use the quantized bosonic field and its functional integration, which is the natural extension of the particle path integral in the case of fields. Let us start with the statistical mechanics of the quantum bosonic field describing interacting identical bosons.

3.2 Complex Gaussian Integrals

In this chapter we will see that for identical bosons described by a quantum field, the imaginary-time Feynman path integral, also called functional integral, can be explicitly calculated in the case of complex Gaussian integrals. A typical complex Gaussian integral is

$$\int_{\mathbb{C}} \frac{dz^* \, dz}{2\pi i} \, e^{-a|z|^2 + bz^* + b^* z} = \frac{1}{a} \, e^{\frac{|b|^2}{a}} \tag{3.8}$$

where z and z^* are the complex integration variables, while a and b are complex parameters. This result is proved by changing variables on the relevant formula

$$\int_{\mathbb{C}} \frac{dz^* \, dz}{2\pi i} \, e^{-|z|^2} = 1 \, . \tag{3.9}$$

The proof, setting $z = r \, e^{i\phi}$ and $z^* = r \, e^{-i\phi}$, goes as follows

$$\begin{aligned} \int_{\mathbb{C}} \frac{dz^* \, dz}{2\pi i} \, e^{-|z|^2} &= \frac{1}{2\pi i} \int_0^{+\infty} dr \; 2i \, r \int_0^{2\pi} d\phi \, e^{-r^2} = \int_0^{+\infty} 2r \, e^{-r^2} \\ &= \left[-e^{-r^2} \right]_0^{+\infty} = 1 \, . \end{aligned} \tag{3.10}$$

For multidimensional complex Gaussian integrals, given the positive definite $N \times N$ hermitian matrix $\hat{A}$ and the complex vector $\mathbf{b}$ with n components, we have

$$\int_{\mathbb{C}^N} \frac{d^N \mathbf{z}^* \, d^N \mathbf{z}}{(2\pi i)^N} \, e^{-\mathbf{z}^* \cdot \hat{A} \mathbf{z} + \mathbf{b} \cdot \mathbf{z}^* + \mathbf{b}^* \cdot \mathbf{z}} = \frac{1}{\det[\hat{A}]} \, e^{\mathbf{b}^* \cdot \hat{A}^{-1} \mathbf{b}} \, . \tag{3.11}$$

Notice that often one uses the formula

$$\frac{1}{\det[\hat{A}]} = e^{\mathrm{Tr}[\ln(\hat{A})]} \tag{3.12}$$

of exponentiation. Here $\det[\hat{A}]$ is the determinant of $\hat{A}$, i.e., the productof the eigenvalues of $\hat{A}$, while $\mathrm{Tr}[\hat{A}]$ is the trace of the matrix $\hat{A}$, i.e., the sum of its eigenvalues.

3.3 Coherent State Path Integral: Single Harmonic Oscillator

The simplest application of the path integral formulation by using complex Gaussian integrals is the harmonic oscillator in the representation of coherent states. The quantum Hamiltonian for a particle with mass m confined in a harmonic potential of frequency ω is given by

$$\hat{H} = \frac{\hat{p}^2}{2m} + \frac{1}{2} m\omega^2 \hat{x}^2 \,. \tag{3.13}$$

It can be rewritten as follows

$$\hat{H} = \hbar\omega \left(\hat{a}^+ \hat{a} + \frac{1}{2} \right) \tag{3.14}$$

in terms of the ladder operators

$$\hat{a} = \sqrt{\frac{m\omega}{2\hbar}} \left(\hat{x} + \frac{i}{m\omega} \hat{p} \right) \tag{3.15}$$

$$\hat{a}^+ = \sqrt{\frac{m\omega}{2\hbar}} \left(\hat{x} - \frac{i}{m\omega} \hat{p} \right) \tag{3.16}$$

The Heisenberg equation of motion of the annihilation operator is given by

$$i\hbar \partial_t \hat{a}(t) = [\hat{a}(t), \hat{H}] = \hbar\omega \, \hat{a}(t) \,. \tag{3.17}$$

Introducing the coherent state $|a(t)\rangle$, that is the eigenstate of the annihilation operator with complex eigenvalue $a(t)$, we immediately find the "classical" equation of motion

$$i\hbar \partial_t a(t) = \hbar\omega \, a(t) \tag{3.18}$$

which is the Euler-Lagrange equation of the following real-time action functional

$$S[a^*, a] = \int_{t_I}^{t_F} dt \left\{ a^*(t) \left(i\hbar\partial_t - \hbar\omega \right) a(t) - \frac{\hbar\omega}{2} \right\} \,. \tag{3.19}$$

Note that we have also included the zero-point energy term, which is washed out by performing the partial derivatives of the Euler-Lagrange equation.

After the Wick rotation $t = -i\tau$ the corresponding Euclidean action reads

$$S_E[a^*, a] = \int_0^{\beta\hbar} d\tau \left\{ a^*(\tau) \left(\hbar\partial_\tau + \hbar\omega \right) a(\tau) + \frac{\hbar\omega}{2} \right\} \,. \tag{3.20}$$

Introducing the Fourier representation of the dynamical variables

$$a(\tau) = \sum_{n=-\infty}^{+\infty} a_n \, e^{-i\omega_n \tau} \tag{3.21}$$

with

$$\omega_n = \frac{2\pi n}{\beta\hbar} \tag{3.22}$$

the bosonic Matsubara frequencies seen in the previous chapter, and observing that

$$\int_0^{\hbar\beta} d\tau \, e^{i(\omega_n - \omega_{n'})\tau} = \hbar\beta \, \delta_{n,n'} \ , \tag{3.23}$$

one finds that the Euclidean action is written in diagonal form as

$$S_E[a^*, a] = \hbar\beta\frac{\hbar\omega}{2} + \hbar \sum_{n=-\infty}^{+\infty} \lambda_n |a_n|^2 \ , \tag{3.24}$$

where

$$\lambda_n = \beta\left(-i\hbar\omega_n + \hbar\omega\right) \ . \tag{3.25}$$

One of the main messages of this section is that the partition function

$$\mathcal{Z}_1 = \mathrm{Tr}[e^{-\beta\hat{H}}] = \mathrm{Tr}[e^{-\beta\hbar\omega(\hat{a}^+\hat{a}+\frac{1}{2})}] \tag{3.26}$$

can be written in terms of coherent states as follows

$$\mathcal{Z}_1 = \int \mathcal{D}[a^*, a] \ \exp\left\{-\frac{S_E[a^*, a]}{\hbar}\right\} \ , \tag{3.27}$$

where

$$\int \mathcal{D}[a^*, a] = \int \prod_\tau \frac{da^*(\tau)da(\tau)}{2\pi i} \tag{3.28}$$

is the measure over the complex numbers $a(\tau)$ and $a^*(\tau)$, which play the role of dynamical variables. This is the so-called coherent state path integral formulation for the single harmonic oscillator.

To prove Eqs. (3.27) and (3.28), we repeat the procedure used in the previous chapter for the quantum particle in a generic potential. Explicitly, we take into account the spectral resolution of the identity

$$\hat{1} = \int_{\mathbb{C}} \frac{da^* da}{2\pi i} \, |a\rangle\langle a| \tag{3.29}$$

discussed in Chapter 1 and also the normalization

$$1 = \int_{\mathbb{C}} \frac{da^* da}{2\pi i} \, e^{-|a|^2} \, . \tag{3.30}$$

Then we write

$$\mathcal{Z}_1 = \text{Tr}[e^{-\beta\hbar\omega(\hat{a}^+\hat{a}+\frac{1}{2})}] = \int_{\mathbb{C}} \frac{da^* da}{2\pi i} \langle a|e^{-\beta\hbar\omega(\hat{a}^+\hat{a}+\frac{1}{2})}|a\rangle \, . \tag{3.31}$$

An approximate, but interesting, consequence is

$$\begin{aligned} \mathcal{Z}_1 &\simeq \int_{\mathbb{C}} \frac{da^* da}{2\pi i} e^{-\beta\hbar\omega\langle a|(\hat{a}^+\hat{a}+\frac{1}{2})|a\rangle} = \int_{\mathbb{C}} \frac{da^* da}{2\pi i} e^{-\beta\hbar\omega(|a|^2+\frac{1}{2})} \\ &= e^{-\beta\frac{\hbar\omega}{2}} \int_{\mathbb{C}} \frac{da^* da}{2\pi i} e^{-\beta\hbar\omega|a|^2} = \frac{e^{-\beta\frac{\hbar\omega}{2}}}{\beta\hbar\omega} \, . \end{aligned} \tag{3.32}$$

We will see that this semiclassical result is reliable in the high-temperature regime, where $\beta\hbar\omega \ll 1$.

To get the complete picture, we must follow some mathematical steps. First, we set $|a_N\rangle$ instead of $|a\rangle$ and also

$$\Delta\tau = \frac{\hbar\beta}{N} \, . \tag{3.33}$$

Second, we apply this formula in the expectation value, namely

$$\begin{aligned} \langle a_N|e^{-\beta\hat{H}}|a_N\rangle &= \langle a_N|e^{-\frac{N\Delta\tau}{\hbar}\hat{H}}|a_N\rangle \\ &= \langle a_N|\left(e^{-\frac{\Delta\tau}{\hbar}\hat{H}}\right)^N|a_N\rangle \\ &= \langle a_N|e^{-\frac{\Delta\tau}{\hbar}\hat{H}} \, ... \, e^{-\frac{\Delta\tau}{\hbar}\hat{H}}|a_N\rangle \, . \end{aligned} \tag{3.34}$$

Third, we insert $N-1$ completeness relations ($j = 1, 2, ..., N-1$)

$$\hat{1} = \int_{\mathbb{C}} \frac{da_j^* da_j}{2\pi i} \, |a_j\rangle\langle a_j| \tag{3.35}$$

in descending order from left to right, each between two exponential operators. In this way we find

$$\int_{\mathbb{C}} \frac{da_N^* da_N}{2\pi i} \langle a_N | e^{-\beta \hat{H}} | a_N \rangle = \int_{\mathbb{C}^N} \left(\prod_{j=1}^{N} \frac{da_j^* da_j}{2\pi i} \right) \prod_{j=1}^{N} \langle a_j | e^{-\frac{\Delta\tau}{\hbar} \hat{H}} | a_{j-1} \rangle \tag{3.36}$$

with the additional constraint $|a_0\rangle = |a_N\rangle$. In the limit $N \to \infty$ one has $\Delta\tau \to 0$ and also

$$\langle a_j | e^{-\frac{\Delta\tau}{\hbar} \hat{H}} | a_{j-1} \rangle = \langle a_j | a_{j-1} \rangle \, e^{-\frac{\Delta\tau}{\hbar} \hbar\omega (a_j^* a_{j-1} + \frac{1}{2})} \,. \tag{3.37}$$

In the first chapter we have seen that the overlap between two coherent states reads

$$\begin{aligned} \langle a_j | a_{j-1} \rangle &= e^{-\frac{1}{2} \left(|a_j|^2 + |a_{j-1}|^2 - 2 a_j^* a_{j-1} \right)} \\ &= e^{-\frac{\Delta\tau}{2} \left(a_j^* \frac{(a_j - a_{j-1})}{\Delta\tau} - \frac{a_{j-1}(a_j^* - a_{j-1}^*)}{\Delta\tau} \right)} \,. \end{aligned} \tag{3.38}$$

Thus, we have

$$\begin{aligned} \mathcal{Z}_1 &= \lim_{N \to \infty} \int_{\mathbb{C}^N} \left(\prod_{j=1}^{N} \frac{da_j^* da_j}{2\pi i} \right) \\ &\quad \times e^{-\frac{1}{\hbar} \sum_{j=1}^{N} \hbar \left(a_j^* \frac{(a_j - a_{j-1})}{2\Delta\tau} - \frac{a_{j-1}(a_j^* - a_{j-1}^*)}{2\Delta\tau} + \omega (a_j^* a_{j-1} + \frac{1}{2}) \right) \Delta\tau} \,. \end{aligned} \tag{3.39}$$

Performing the Gaussian integrals in Eq. (3.39), after some manipulations we get

$$\mathcal{Z}_1 = e^{-\beta\hbar\omega/2} \lim_{N \to \infty} \left(\left[1 - \prod_{j=1}^{N} (1 - \frac{\beta\hbar\omega}{N}) \right] \right)^{-1} = e^{-\beta\hbar\omega/2} \frac{1}{1 - e^{-\beta\hbar\omega}} \tag{3.40}$$

that is the expected exact result, found in the Chapter 2 by using the path integral but without coherent states.

It is extremely important to stress that, setting $\tau = j\Delta\tau$ and $a(\tau) = a_j$ in Eq. (3.39), in the continuum limit we formally obtain Eqs. (3.27) and (3.28). Quite remarkably, also working in the continuum, i.e., using Eqs. (3.11), (3.24), (3.27), and (3.28), one can study the thermodynamics of the harmonic oscillator. There is, however, an important issue related to the time ordering because, strictly speaking, in the continuum the Euclidean action is

$$S_E[a^*,a] = \int_0^{\beta\hbar} d\tau \left\{ a^*(\tau+0^+)\left(\hbar\partial_\tau + \hbar\omega\right)a(\tau) + \frac{\hbar\omega}{2} \right\} \tag{3.41}$$

$$= \hbar\beta\frac{\hbar\omega}{2} + \hbar \sum_{n=-\infty}^{+\infty} \lambda_n e^{i\omega_n 0^+} |a_n|^2 \,, \tag{3.42}$$

where 0^+ remembers that $a^*(\tau+0^+)$ is calculated at the later time with respect to $a(\tau)$. Thus, the partition function is given by

$$\begin{aligned}
\mathcal{Z}_1 &= \int \mathcal{D}[\{a_n^*\},\{a_n\}] \, \exp\left\{ -\beta\frac{\hbar\omega}{2} - \sum_n \lambda_n e^{i\omega_n 0^+} |a_n|^2 \right\} \\
&= e^{-\beta\hbar\omega/2} \int \prod_{n=-\infty}^{+\infty} \frac{da_n^* \, da_n}{(2\pi i)} \, \exp\left\{ -\sum_n \lambda_n e^{i\omega_n 0^+} |a_n|^2 \right\} \\
&= e^{-\beta\hbar\omega/2} \left(\prod_{n=-\infty}^{+\infty} \lambda_n e^{i\omega_n 0^+} \right)^{-1} \\
&= e^{-\beta\hbar\omega/2} \left(\prod_{n=-\infty}^{+\infty} \beta(-i\hbar\omega_n + \hbar\omega) e^{i\omega_n 0^+} \right)^{-1} \\
&= e^{-\beta\hbar\omega/2} \left(\prod_{n=-\infty}^{+\infty} (-i2\pi\, n + \beta\hbar\omega) e^{i\omega_n 0^+} \right)^{-1} \\
&= \frac{e^{-\beta\hbar\omega/2}}{1 - e^{-\beta\hbar\omega}} \,.
\end{aligned} \tag{3.43}$$

That is again the expected exact formula. In this way we prove that the use of Eq. (3.27) is justified. In the high-temperature regime $\beta\hbar\omega \gg 1$, i.e., $k_B T \gg \hbar\omega$, Eq. (3.43) reduces to Eq. (3.32). Notice that Eq. (3.43) is based on the formula

$$\prod_{n=-\infty}^{+\infty} (-i2\pi\, n + x)\, e^{i\omega_n 0^+} = 1 - e^{-x} \,, \tag{3.44}$$

whose detailed derivation is discussed in Problem 3.4. In Problem 3.5 we instead calculate the infinite product without the use of the convergence factor, finding a slightly different result, which contains also the zero-point energy.

Concluding this subsection we underline that, following the procedure discussed here but working in real time, one can calculate the propagator

$$K_{\text{harm}}(a_F, t_F|a_F, t_F) = \langle a_F t_F | a_I t_I \rangle = \langle a_F | e^{-\frac{i}{\hbar}\hbar\omega(\hat{a}^+\hat{a}+\frac{1}{2})(t_F - t_I)} | a_I \rangle \tag{3.45}$$

that gives the transition amplitude between the time-dependent coherent state $|a_I t_I\rangle = |a_I(t_I)\rangle$ and the time-dependent coherent state $|a_F t_F\rangle = |a_F(t_F)\rangle$. The final result reads

$$K_{\text{harm}}(a_F, t_F|a_F, t_F) = e^{-\frac{1}{2}|a_F|^2}\, e^{-\frac{1}{2}|a_I|^2}\, e^{a_F^* a_I e^{-i\omega(t_F - t_I)}}\, e^{-i\frac{\omega}{2}(t_F - t_I)} \tag{3.46}$$

that is a quite peculiar expression. Eq. (3.46) can also be derived immediately from the formula (1.205) of the superposition of coherent states discussed in Chapter 1, remembering that $a_F(t_F) = a_F(0)\, e^{-i\omega t_F}$ and $a_I(t_I) = a_I(0)\, e^{-i\omega t_I}$. Just to close the circle, having the propagator (3.46), the partition function can be obtained as

$$\begin{aligned}
\mathcal{Z}_1 &= \int_{\mathbb{C}} \frac{da^* da}{2\pi i} K_{\text{harm}}(a, -i\hbar\beta|a, 0) \\
&= e^{-\beta\hbar\omega/2} \int_{\mathbb{C}} \frac{da^* da}{2\pi i}\, e^{-|a|^2(1-e^{-\beta\hbar\omega})} \\
&= e^{-\beta\hbar\omega/2} \int_0^{+\infty} dr\, 2r\, e^{-r^2(1-e^{-\beta\hbar\omega})} \\
&= e^{-\beta\hbar\omega/2} \int_0^{+\infty} dy\, e^{-y(1-e^{-\beta\hbar\omega})} \\
&= \frac{e^{-\beta\hbar\omega/2}}{1 - e^{-\beta\hbar\omega}} \, .
\end{aligned} \tag{3.47}$$

A last comment is dutiful. The partition function $\mathcal{Z}_1$ of the single harmonic oscillator can be much easier deduced without invoking a path integral formalism, but instead using the Fock states $|n\rangle$, that are eigenstates of the number operator $\hat{N} = \hat{a}^+\hat{a}$

with eigenvalues $n = 0, 1, 2, ...$ Specifically

$$
\begin{aligned}
\mathcal{Z}_1 &= Tr[e^{-\beta\hbar\omega(\hat{N}+\frac{1}{2})}] = \sum_{n=0}^{\infty} \langle n| e^{-\beta\hbar\omega(\hat{N}+\frac{1}{2})} |n\rangle = \sum_{n=0}^{\infty} e^{-\beta\hbar\omega(n+\frac{1}{2})} \\
&= e^{-\beta\hbar\omega/2} \sum_{n=0}^{\infty} \left(e^{-\beta\hbar\omega}\right)^n = e^{-\beta\hbar\omega/2} \frac{1}{1-e^{-\beta\hbar\omega}}
\end{aligned}
\tag{3.48}
$$

where in the last step, we have used the geometric series $\sum_{n=0}^{\infty} x^n = 1/(1-x)$ for $|x| < 1$.

3.4 Partition Function of Interacting Bosons

For a system of non-relativistic interacting bosons the partition function is given by

$$
\mathcal{Z} = \mathrm{Tr}[e^{-\beta(\hat{H}-\mu\hat{N})}] , \tag{3.49}
$$

where the second-quantized Hamiltonian is

$$
\begin{aligned}
\hat{H} = &\int d^3\mathbf{r}\ \hat{\psi}^+(\mathbf{r}) \Big(-\frac{\hbar^2}{2m}\nabla^2 + U(\mathbf{r}) \Big) \hat{\psi}(\mathbf{r}) \\
&+ \frac{1}{2} \int d^3\mathbf{r}\ d^3\mathbf{r}'\ \hat{\psi}^+(\mathbf{r})\ \hat{\psi}^+(\mathbf{r}')\ V(\mathbf{r},\mathbf{r}')\ \hat{\psi}(\mathbf{r}')\ \hat{\psi}(\mathbf{r}) ,
\end{aligned}
\tag{3.50}
$$

with $U(\mathbf{r})$ the external potential acting on spinless bosons described by the field operator $\hat{\psi}(\mathbf{r})$, $V(\mathbf{r},\mathbf{r}')$ the inter-particle interaction, and the number operator reads

$$
\hat{N} = \int d^3\mathbf{r}\ \hat{\psi}^+(\mathbf{r})\hat{\psi}(\mathbf{r}) . \tag{3.51}
$$

In Eq. (3.49), as usual, $\beta = 1/(k_B T)$ with k_B the Boltzmann constant and T the temperature, while μ is the chemical potential. In the first chapter we have seen that, by expanding the field operator $\hat{\psi}(\mathbf{r})$ as

$$
\hat{\psi}(\mathbf{r}) = \sum_{\alpha} \hat{c}_{\alpha}\, \phi_{\alpha}(\mathbf{r}) , \tag{3.52}
$$

where $\phi_\alpha(\mathbf{r})$ are the orthonormal eigenfunctions of the non-interacting problems with single-particle energies ϵ_α, one finds

$$\hat{H} = \sum_\alpha \epsilon_\alpha \; \hat{c}_\alpha^+ \hat{c}_\alpha + \sum_{\alpha\beta\gamma\delta} V_{\alpha\beta\gamma\delta} \; \hat{c}_\alpha^+ \hat{c}_\beta^+ \hat{c}_\delta \hat{c}_\gamma \; , \qquad (3.53)$$

where

$$V_{\alpha\beta\delta\gamma} = \int d^3\mathbf{r} \; d^3\mathbf{r}' \; \phi_\alpha^*(\mathbf{r}) \; \phi_\beta^*(\mathbf{r}') \; V(\mathbf{r}, \mathbf{r}') \; \phi_\delta(\mathbf{r}') \; \phi_\gamma(\mathbf{r}) \; , \qquad (3.54)$$

and also

$$\hat{N} = \sum_\alpha \hat{c}_\alpha^+ \hat{c}_\alpha \; . \qquad (3.55)$$

Moreover, we have seen that the generalized coherent state

$$|\psi\rangle = \prod_\alpha |c_\alpha\rangle = e^{-\sum_\alpha |c_\alpha|^2/2} \, e^{\sum_\alpha c_\alpha \hat{c}_\alpha^+} |0\rangle \; , \qquad (3.56)$$

is such that

$$\hat{\psi}(\mathbf{r}) \, |\psi\rangle = \psi(\mathbf{r}) \, |\psi\rangle \; , \qquad (3.57)$$

where

$$\psi(\mathbf{r}) = \sum_\alpha c_\alpha \, \phi_\alpha(\mathbf{r}) \qquad (3.58)$$

is the "classical" Schrödinger field with c_α the complex eigenvalues of $\hat{c}_\alpha$. Clearly, it follows that

$$\hat{c}_\alpha |\psi\rangle = c_\alpha |\psi\rangle \qquad \text{for any } \alpha \; . \qquad (3.59)$$

Note that $|\psi\rangle$ is the tensor product of all single-mode coherent states $|c_\alpha\rangle$. Due to their normalization we get

$$\langle\psi|\psi\rangle = 1 \; . \qquad (3.60)$$

The overlap between two generalized coherent states $|\psi\rangle$ and $|\tilde{\psi}\rangle$ is instead given by

$$\langle\psi|\tilde{\psi}\rangle = e^{-\frac{1}{2}\sum_\alpha [c_\alpha^*(c_\alpha - \tilde{c}_\alpha) - (c_\alpha^* - \tilde{c}_\alpha^*)\tilde{c}_\alpha]} \; . \qquad (3.61)$$

In Chapter 1 and also in the previous section, we have seen that the single-mode coherent state $|c_\alpha\rangle$ satisfies the remarkable completeness relation

$$\int \frac{dc_\alpha^* dc_\alpha}{2\pi i} |c_\alpha\rangle\langle c_\alpha| = \hat{1} . \tag{3.62}$$

It is possible to prove that the set of single-mode coherent states $\{|c_\alpha\rangle\}$ obtained by varying α is over-complete and the generalized coherent state $|\psi\rangle$ satisfies the completeness relation

$$\int d[\psi^*, \psi] \, |\psi\rangle\langle\psi| = \hat{1} , \tag{3.63}$$

where the integration measure is defined by

$$\int d[\psi^*, \psi] = \int \prod_\alpha \frac{dc_\alpha^* dc_\alpha}{2\pi i} \tag{3.64}$$

with $\psi^*(\mathbf{r})$ the complex conjugate of $\psi(\mathbf{r})$ such that

$$\langle\psi| \, \hat{\psi}^+(\mathbf{r}) = \psi^*(\mathbf{r}) \, \langle\psi| . \tag{3.65}$$

Notice that c_α and $\psi(\mathbf{r})$ represent the same complex field in reciprocal spaces, and the integration measure can also be written as

$$\int d[\psi^*, \psi] = \int \prod_{\mathbf{r}} \frac{d\psi^*(\mathbf{r}) d\psi(\mathbf{r})}{2\pi i} . \tag{3.66}$$

Moreover, Eq. (3.61) can also be written as

$$\langle\psi|\tilde{\psi}\rangle = e^{-\frac{1}{2}\int d^3\mathbf{r}[\psi^*(\mathbf{r})(\psi(\mathbf{r})-\tilde{\psi}(\mathbf{r}))-(\psi^*(\mathbf{r})-\tilde{\psi}^*(\mathbf{r}))\tilde{\psi}(\mathbf{r})]} . \tag{3.67}$$

The key point is that the partition function $\mathcal{Z}$ can be expressed as a functional integral taking into account that the trace of an operator can be performed by using any set of basis states, and, in particular, the basis of generalized coherent states. Thus, we write

$$\mathcal{Z} = \mathrm{Tr}[e^{-\beta(\hat{H}-\mu\hat{N})}] = \int d[\psi^*, \psi] \, \langle\psi|e^{-\beta(\hat{H}-\mu\hat{N})}|\psi\rangle , \tag{3.68}$$

where $|\psi\rangle$ is the generalized coherent state.

3.4.1 Semiclassical Approximation and Imaginary Time

The main problem in the functional representation of the partition function $\mathcal{Z}$ is the calculation of

$$\langle\psi|e^{-\beta(\hat{H}-\mu\hat{N})}|\psi\rangle \,. \tag{3.69}$$

The simple semiclassical approximation

$$\langle\psi|e^{-\beta(\hat{H}-\mu\hat{N})}|\psi\rangle \simeq e^{-\beta\langle\psi|(\hat{H}-\mu\hat{N})|\psi\rangle} \tag{3.70}$$

gives

$$\mathcal{Z} = \int d[\psi^*,\psi]\, e^{-\beta(E[\psi^*,\psi]-\mu N[\psi^*,\psi])} \,, \tag{3.71}$$

where

$$\begin{aligned} E[\psi,\psi^*] &= \int d^3\mathbf{r}\ \psi^*(\mathbf{r})\Big(-\frac{\hbar^2}{2m}\nabla^2 + U(\mathbf{r})\Big)\psi(\mathbf{r}) \\ &+ \frac{1}{2}\int d^3\mathbf{r}\, d^3\mathbf{r}'\, |\psi(\mathbf{r})|^2\, V(\mathbf{r},\mathbf{r}')\, |\psi(\mathbf{r}')|^2 \end{aligned} \tag{3.72}$$

and

$$N[\psi,\psi^*] = \int d^3\mathbf{r}\, \psi^*(\mathbf{r})\psi(\mathbf{r}) \,. \tag{3.73}$$

The partition function given by Eq. (3.71) is called semiclassical because it is usually reliable only at high temperatures, where quantum statistics plays a marginal role.

To better treat the expectation value of Eq. (3.69), we first write $|\psi_0\rangle$ instead of $|\psi\rangle$ and also (with $M \to \infty$)

$$\Delta\tau = \frac{\hbar\beta}{M} \,. \tag{3.74}$$

As we shall see immediately, this number $\Delta\tau$ plays the role of the step interval of an imaginary time τ within a Wick rotation framework. In fact, we obtain

$$\begin{aligned} \langle\psi_0|e^{-\beta(\hat{H}-\mu\hat{N})}|\psi_0\rangle &= \langle\psi_0|e^{-M\Delta\tau(\hat{H}-\mu\hat{N})/\hbar}|\psi_0\rangle \\ &= \langle\psi_0|\left(e^{-\Delta\tau(\hat{H}-\mu\hat{N})/\hbar}\right)^M|\psi_0\rangle \\ &= \langle\psi_0|e^{-\Delta\tau(\hat{H}-\mu\hat{N})/\hbar}\ldots e^{-\Delta\tau(\hat{H}-\mu\hat{N})/\hbar}|\psi_0\rangle \,. \end{aligned} \tag{3.75}$$

Then we insert $M-1$ completeness relations for the general coherent state $|\psi_j\rangle$ $(j=1,...,M-1)$

$$\int d[\psi_j,\psi_j^*]\,|\psi_j\rangle\langle\psi_j| = 1 \tag{3.76}$$

to get

$$\langle\psi_0|e^{-\beta(\hat{H}-\mu\hat{N})}|\psi_0\rangle = \int \left(\prod_{j=1}^{M-1} d[\psi_j^*,\psi_j]\right)\prod_{j=1}^{M}\langle\psi_j|e^{-\Delta\tau(\hat{H}-\mu\hat{N})/\hbar}|\psi_{j-1}\rangle \tag{3.77}$$

imposing that $|\psi_0\rangle = |\psi_M\rangle$. In the limit $M\to\infty$ one has $\Delta\tau\to 0$ and also

$$\langle\psi_j|e^{-\Delta\tau(\hat{H}-\mu\hat{N})/\hbar}|\psi_{j-1}\rangle = \langle\psi_j|\psi_{j-1}\rangle\, e^{-\Delta\tau(E[\psi_j^*,\psi_j]-\mu N[\psi_j^*,\psi_j])/\hbar}\,. \tag{3.78}$$

Moreover, from Eq. (3.67) the overlap between two general coherent states reads

$$\langle\psi_j|\psi_{j-1}\rangle = e^{-\frac{\Delta\tau}{2}\int d^3\mathbf{r}\left(\psi_j^*(\mathbf{r})\frac{[\psi_j(\mathbf{r})-\psi_{j-1}(\mathbf{r})]}{\Delta\tau}-\psi_j(\mathbf{r})\frac{[\psi_j^*(\mathbf{r})-\psi_{j-1}^*(\mathbf{r})]}{\Delta\tau}\right)}\,. \tag{3.79}$$

Finally, setting $\tau=j\Delta\tau$, $\psi(\mathbf{r},\tau)=\psi_j(\mathbf{r})$ and

$$\frac{\partial}{\partial\tau}\psi(\mathbf{r},\tau) = \frac{[\psi_j(\mathbf{r})-\psi_{j-1}(\mathbf{r})]}{\Delta\tau}\,, \tag{3.80}$$

we obtain the partition function

$$\mathcal{Z} = \int \mathcal{D}[\psi,\psi^*]\;\exp\left\{-\frac{S_E[\psi,\psi^*]}{\hbar}\right\}\,, \tag{3.81}$$

where

$$S_E[\psi,\psi^*] = \int_0^{\hbar\beta} d\tau \int d^3\mathbf{r}\,\mathscr{L}_E(\psi^*,\psi) \tag{3.82}$$

is the Euclidean (i.e., with imaginary time) action of the Euclidean Lagrangian density

$$\begin{aligned}\mathscr{L}_E(\psi^*,\psi) &= \frac{\hbar}{2}\Big(\psi^*(\mathbf{r},\tau)\frac{\partial}{\partial\tau}\psi(\mathbf{r},\tau)-\psi(\mathbf{r},\tau)\frac{\partial}{\partial\tau}\psi^*(\mathbf{r},\tau)\Big)\\ &+ \psi^*(\mathbf{r},\tau)\Big(-\frac{\hbar^2}{2m}\nabla^2+U(\mathbf{r})-\mu\Big)\psi(\mathbf{r},\tau) \\ &+ \frac{1}{2}\int d^3\mathbf{r}'\;|\psi(\mathbf{r}',\tau)|^2\,V(\mathbf{r},\mathbf{r}')\,|\psi(\mathbf{r},\tau)|^2\end{aligned} \tag{3.83}$$

with time-periodic boundary conditions $\psi(\mathbf{r},0) = \psi(\mathbf{r},\hbar\beta)$ and

$$\int \mathcal{D}[\psi,\psi^*] = \int \prod_{(\mathbf{r},\tau)} \frac{d\psi(\mathbf{r},\tau)\, d\psi^*(\mathbf{r},\tau)}{2\pi i} \,. \tag{3.84}$$

We stress again that a semiclassical treatment of the expectation value (3.69) gives a semiclassical partition function (3.71), which is written as a functional integration of the time-independent energy functional. Instead, the exact treatment of Eq. (3.69) produces the exact partition function (3.81), which is expressed as a functional integration of the imaginary-time action functional. It is important to emphasize that, treating this bosonic problem of identical particles, we have not included the (divergent) vacuum energy nor in the quantum many-body Hamiltonian $\hat{H}$ nor in the corresponding "classical" action functional S_E.

3.4.2 Non-Interacting Bosons in a Generic Trapping Potential

In the absence of inter-particle interaction, the Euclidean Lagrangian density can be written in a simplified form as

$$\mathscr{L}_E = \psi^*(\mathbf{r},\tau)\Big(\hbar\frac{\partial}{\partial\tau} - \frac{\hbar^2}{2m}\nabla^2 + U(\mathbf{r}) - \mu\Big)\psi(\mathbf{r},\tau) \,. \tag{3.85}$$

The field $\psi(\mathbf{r},\tau)$ can be expanded in the basis $\phi_\alpha(\mathbf{r})$ of the eigenfunctions of the stationry Schrödinger problem

$$\left(-\frac{\hbar^2}{2m}\nabla^2 + U(\mathbf{r})\right)\phi_\alpha(\mathbf{r}) = \epsilon_\alpha\, \phi_\alpha(\mathbf{r}) \tag{3.86}$$

where α labels the set of possible single-particle quantum numbers of the problem, and ϵ_α is the generic eigenvalue of the single-particle energy. The expansion of $\psi(\mathbf{r},\tau)$ reads

$$\psi(\mathbf{r},\tau) = \sum_\alpha c_\alpha(\tau)\, \phi_\alpha(\mathbf{r}) \tag{3.87}$$

where $c_\alpha(\tau)$ is the generic time-dependent coefficient of this expansion. Imposing the ortho-normalization

$$\int d^3\mathbf{r}\, \phi^*_\alpha(\mathbf{r})\phi_{\alpha'}(\mathbf{r}) = \delta_{\alpha,\alpha'} \tag{3.88}$$

and inserting the expansion (3.87) into the Euclidean Lagrangian (3.85), we get

$$\mathscr{L}_E = \sum_\alpha c^*_\alpha(\tau)\Big(\hbar\frac{\partial}{\partial\tau} + \epsilon_\alpha - \mu\Big)c_\alpha(\tau) \,. \tag{3.89}$$

Moreover, we can also expand in a Fourier basis the complex coefficients $c_\alpha(\tau)$, namely

$$c_\alpha(\tau) = \sum_{n=-\infty}^{+\infty} c_{\alpha,n}\, e^{-i\omega_n\tau} \tag{3.90}$$

where, again, $\omega_n = 2\pi n/(\hbar\beta)$ are the bosonic Matsubara frequencies. In this way the Euclidean action functional becomes

$$S_E = \hbar\beta\sum_\alpha\sum_{n=-\infty}^{+\infty} c^*_{\alpha,n}\Big(-i\hbar\omega_n + \epsilon_\alpha - \mu\Big)c_{\alpha,n} \tag{3.91}$$

and the corresponding partition function is

$$\mathcal{Z} = \int \mathcal{D}[\{c^*_{\alpha,n}\}\{c_{\alpha,n}\}]\, e^{-\frac{1}{\hbar}S_E[\{c^*_{\alpha,n}\},\{c_{\alpha,n}\}]} \,. \tag{3.92}$$

It is quite clear that this partition function the nothing else than the infinite product of the harmonic oscillator partition functions with charactristic frequencies $\omega_\alpha = (\epsilon_\alpha - \mu)/\hbar$. Thus, we have

$$\mathcal{Z} = \prod_\alpha \frac{1}{1 - e^{-\beta\xi_\alpha}} \,, \tag{3.93}$$

where

$$\xi_\alpha = \epsilon_\alpha - \mu \tag{3.94}$$

is the shifted single-particle energy. The grand potential is then

$$\Omega = -\frac{1}{\beta}\ln(\mathcal{Z}) = \frac{1}{\beta}\sum_\alpha \ln\big(1 - e^{\beta\xi_\alpha}\big) \,. \tag{3.95}$$

The average number of particles can be obtained by using the formula

$$N = \langle\hat{N}\rangle = -\left(\frac{\partial\Omega}{\partial\mu}\right)_T \tag{3.96}$$

finding

$$N = \sum_\alpha \frac{1}{e^{\beta\xi_\alpha} - 1} \,. \tag{3.97}$$

3.4.3 Ideal and Uniform Case: Bose-Einstein Condensation

First, we analyze the case of non-interacting bosons with a vanishing external potential, namely

$$U(\mathbf{r}) = 0 \ , \tag{3.98}$$

where exact analytical results can be obtained in any spatial dimension D. In this case the Lagrangian density reads

$$\mathscr{L}_E = \psi^*(\mathbf{r},\tau)\left[\hbar\frac{\partial}{\partial\tau} - \frac{\hbar^2}{2m}\nabla^2 - \mu\right]\psi(\mathbf{r},\tau) \ . \tag{3.99}$$

Introducing the Fourier representation of the fluctuation field

$$\psi(\mathbf{r},\tau) = \frac{1}{L^{D/2}}\sum_{n=-\infty}^{+\infty}\sum_{\mathbf{k}} c_{\mathbf{k},n}\ e^{i(\mathbf{k}\cdot\mathbf{r}-\omega_n\tau)} \ , \tag{3.100}$$

where $\mathbf{k}$ is the D-dimensional wavevector and $\omega_n = 2\pi n/(\hbar\beta)$ and the bosonic Matsubara frequencies, the Euclidean action of the ideal Bose gas can be written in a diagonal form as

$$S_E[\psi^*,\psi] \ = \ \sum_{n=-\infty}^{+\infty}\sum_{\mathbf{k}} \hbar\lambda_{\mathbf{k},n}\,|c_{\mathbf{k},n}|^2 \tag{3.101}$$

with

$$\lambda_{\mathbf{k},n} = \beta\left(-i\hbar\omega_n + \frac{\hbar^2k^2}{2m} - \mu\right) \ . \tag{3.102}$$

The partition function is given by

$$\begin{aligned}\mathcal{Z} \ &= \ \int\mathcal{D}[c^*_{\mathbf{k},n},c_{\mathbf{k},n}]\exp\left\{-\sum_{+\infty}^{+\infty}\sum_{\mathbf{k}}\lambda_{\mathbf{k},n}|c_{\mathbf{k},n}|^2\right\} = \frac{1}{\prod_{n=-\infty}^{+\infty}\prod_{\mathbf{k}}\lambda_{\mathbf{k}n}} \\ &= \ \frac{1}{\prod_{\mathbf{k}}\prod_{n=-\infty}^{+\infty}\left(-i2\pi n + \beta(\frac{\hbar^2k^2}{2m} - \mu)\right)} = \frac{1}{\prod_{\mathbf{k}}(1-e^{-\beta\xi_k})}\end{aligned} \tag{3.103}$$

where ξ_k is the shifted free-particle spectrum, i.e.

$$\xi_k = \frac{\hbar^2k^2}{2m} - \mu \ . \tag{3.104}$$

It follows that the corresponding grand potential reads

$$\Omega = \frac{1}{\beta} \sum_{\mathbf{k}} \ln\left(1 - e^{-\beta\xi_k}\right) . \tag{3.105}$$

As previously seen, the number density $n = N/L^D$ is obtained from the thermodynamic relation

$$n = -\frac{1}{L^D} \left(\frac{\partial \Omega}{\partial \mu} \right)_{T,L^D} , \tag{3.106}$$

which gives

$$n = \frac{1}{L^D} \sum_{\mathbf{k}} \frac{1}{e^{\beta\xi_k} - 1} = \sum_{\mathbf{k}} n_{\mathbf{k}} . \tag{3.107}$$

In the continuum limit $\sum_{\mathbf{k}} \to L^D \int d^D\mathbf{k}/(2\pi)^D$, by setting $\mu = 0$ (Bose-condensed phase) and explicitly writing $n_{\mathbf{0}}$ as the number density of bosons with wavenumber $\mathbf{k} = \mathbf{0}$ from Eq. (3.107) one gets

$$n = n_{\mathbf{0}} + \int \frac{d^D\mathbf{k}}{(2\pi)^D} \frac{1}{e^{\frac{\hbar^2 k^2}{2mk_B T}} - 1} , \tag{3.108}$$

which indeed gives the Bose-Einstein condensate density $n_{\mathbf{0}}$ as a function of the temperature T at fixed total number density n.

The critical temperature T_{BEC} of Bose-Einstein condensation is obtained by setting $n_0 = 0$ in the previous equation. In this way one finds

$$k_B T_{\text{BEC}} = \begin{cases} \frac{1}{2\pi\zeta(3/2)^{2/3}} \frac{\hbar^2}{m} n^{2/3} & \text{for } D = 3 \\ 0 & \text{for } D = 2 \\ \text{no solution} & \text{for } D = 1 \end{cases} \tag{3.109}$$

where $\zeta(x)$ is the Riemann zeta function and $\zeta(3/2) = 2.6123$. This result was extended to interacting systems by David Mermin and Herbert Wagner in 1966. The Mermin-Wagner theorem states that there is no Bose-Einstein condensation at finite temperature in homogeneous systems with sufficiently short-range interactions in dimensions $D \leq 2$.

In the three-dimensional case ($D = 3$) from the equation

$$n = n_0 + \int \frac{d^3\mathbf{k}}{(2\pi)^3} \frac{1}{e^{\frac{\hbar^2 k^2}{2mk_BT}} - 1} , \tag{3.110}$$

we explicitly obtain

$$n = n_0 + \zeta(3/2) \left(\frac{mk_BT}{2\pi\hbar^2} \right)^{3/2} . \tag{3.111}$$

It follows that

$$\frac{n_0}{n} = 1 - \frac{\zeta(3/2)\left(\frac{mk_B}{2\pi\hbar^2}\right)^{3/2} T^{3/2}}{n} = 1 - \frac{\zeta(3/2)\left(\frac{mk_B}{2\pi\hbar^2}\right)^{3/2} T^{3/2}}{\zeta(3/2)\left(\frac{mk_B}{2\pi\hbar^2}\right)^{3/2} T_{\text{BEC}}^{3/2}} . \tag{3.112}$$

Thus, the condensate fraction reads

$$\frac{n_0}{n} = 1 - \left(\frac{T}{T_{\text{BEC}}} \right)^{3/2} . \tag{3.113}$$

The critical temperature of Eq. (3.109) and this formula for the condensate fraction of non-interacting massive bosons were obtained in 1924 by Albert Einstein, extending previous results derived by Satyendra Nath Bose for a gas of photons (massless bosons).

3.5 Ideal Bose Gas in a Harmonic Trap

Let us consider a bosonic gas of N non-interacting spinless particles under harmonic confinement given by the external trapping potential

$$U(\mathbf{r}) = \frac{1}{2} m\omega^2 (x^2 + y^2 + z^2) . \tag{3.114}$$

This is a typical problem one faces in the case of ultracold atoms dilute gases made of alkali-metal atoms, where the trapping potential can originate from the presence of a spatially-dependent magnetic field coupled to the dipole magnetic moment of each atom.

We want to calculate the critical temperature $T_{\rm BEC}$ of Bose-Einstein condensation as a function of the number N of bosons and the condensate fraction N_0/N as a function of the temperature T. To reach this goal, one must generalize Eq. (3.107), taking into account the external potential $U(\mathbf{r})$ acting on bosons. This can be done by introducing the shifted single-particle energy

$$\xi_q(\mathbf{r}) = \frac{\hbar^2 k^2}{2m} + U(\mathbf{r}) - \mu \,. \tag{3.115}$$

In this way one can write the local number density as

$$n(\mathbf{r}) = n_0(\mathbf{r}) + \int \frac{d^3\mathbf{k}}{(2\pi)^3} \frac{1}{e^{\frac{\xi_q(\mathbf{r})}{k_B T}} - 1} \,, \tag{3.116}$$

where $n_0(\mathbf{r})$ is the local condensate density. Below $T_{\rm BEC}$ one can set $\mu = 0$. Moreover, at $T_{\rm BEC}$ by definition $n_0(\mathbf{r}) = 0$ and consequently

$$n(\mathbf{r}) = \int \frac{d^3\mathbf{k}}{(2\pi)^3} \frac{1}{e^{\frac{\frac{\hbar^2 k^2}{2m} + U(\mathbf{r})}{k_B T_{\rm BEC}}} - 1} \,. \tag{3.117}$$

After integration over momenta, one obtains

$$n(\mathbf{r}) = \frac{1}{\lambda_{T_{\rm BEC}}^3} g_{3/2}(e^{-\frac{U(\mathbf{r})}{k_B T_{\rm BEC}}}) \,, \tag{3.118}$$

where

$$\lambda_{T_{\rm BEC}} = \left(\frac{2\pi\hbar^2}{m k_B T_{\rm BEC}} \right)^{1/2} \tag{3.119}$$

is the de Broglie wavelength at the critical temperature $T_{\rm BEC}$ and $g_n(x)$ is the Bose function, defined as

$$g_n(x) = \frac{1}{\Gamma(n)} \int_0^{+\infty} dy \, \frac{x \, y^{n-1} \, e^{-y}}{1 - x \, e^{-y}} = \sum_{i=1}^{+\infty} \frac{x^i}{i^n} \,, \tag{3.120}$$

with

$$\Gamma(z) = \int_0^{+\infty} dt \, t^{z-1} \, e^{-t} \tag{3.121}$$

the Euler gamma function, and the last equality holds only if $|x| < 1$. Moreover, integrating over coordinates, one gets

$$N = \int d^3\mathbf{r}\; n(\mathbf{r}) = g_3(1) \left(\frac{\hbar\omega}{k_B T_{\rm BEC}} \right)^3 , \tag{3.122}$$

from which

$$T_{\rm BEC} = \frac{\hbar\omega}{g_3(1)\; k_B} N^{1/3} . \tag{3.123}$$

Notice that $g_3(1) = \zeta(3) = 1.2021$. For $T \leq T_{\rm BEC}$ one can write

$$N = N_0 + \int \frac{d^3\mathbf{r} d^3\mathbf{k}}{(2\pi)^3} \frac{1}{e^{\frac{\frac{\hbar^2 k^2}{2m} + U(\mathbf{r})}{k_B T}} - 1} \tag{3.124}$$

and consequently

$$\frac{N_0}{N} = 1 - \frac{1}{N} \int \frac{d^3\mathbf{r} d^3\mathbf{k}}{(2\pi)^3} \frac{1}{e^{\frac{\frac{\hbar^2 k^2}{2m} + U(\mathbf{r})}{k_B T}} - 1} . \tag{3.125}$$

After integration over coordinates and momenta, and taking into account the formula of $T_{\rm BEC}$, one finally gets

$$\frac{N_0}{N} = 1 - \left(\frac{T}{T_{\rm BEC}} \right)^3 . \tag{3.126}$$

3.6 Basic Notions of Quantum Scattering

In the quantum theory of scattering, one usually assumes that an incoming particle, described by a plane wave along the z-axis, scatters with a target, modeled by an external spherically-symmetric potential $V(\mathbf{r})$. The same result is obtained by considering the scattering between two identical particles of mass m in the center of mass frame. In this case the effective particle has the reduced mass $m_r = m/2$, and $V(\mathbf{r})$ represents the inter-particle potential between the two particles with $\mathbf{r}$ the relative position vector. In the following we will consider the collision between two identical particles which are described by the stationary Schrödinger equation

$$\left(-\frac{\hbar^2}{2m_r}\nabla^2 + V(\mathbf{r})\right)\psi(\mathbf{r}) = E_k\,\psi(\mathbf{r}) \tag{3.127}$$

with $E_k = \hbar^2 k^2/(2m_r)$ the energy of the effective particle characterized by the initial wavevector $\mathbf{k} = (0,0,k)$. An asymptotic solution of Eq. (3.127) reads

$$\psi(\mathbf{r}) \simeq e^{ikz} + f(k,\theta)\frac{e^{ikr}}{r} , \tag{3.128}$$

which is a superposition between the initial plane wave e^{ikz} along the z axis and the final spherical wave e^{ikr}/r, with $f(k,\theta)$ is the so-called scattering amplitude. Correspondingly, the differential cross section in the infinitesimal solid angle $d\Omega$ is given by

$$\frac{d\sigma}{d\Omega} = |f(k,\theta)|^2 , \tag{3.129}$$

which is the scattering probability per unit time.

The scattering amplitude $f(k,\theta)$ is often written as

$$f(k,\theta) = \sum_{l=0}^{\infty}(2l+1) f_l(k)\, P_l(\cos(\theta)) , \tag{3.130}$$

where $P_l(x)$ are the Legendre polynomials and $f_l(k)$ are the partial-wave scattering amplitudes. Moreover, usually one introduces the phase shift $\delta_l(k)$ such that

$$f_l(k) = \frac{e^{i\delta_l(k)}}{k}\,\sin(\delta_l(k)) . \tag{3.131}$$

The corresponding integrated cross section $\sigma(k)$ reads

$$\begin{aligned}\sigma(k) &= \int \frac{d\sigma}{d\Omega}\,d\Omega = \int_0^{2\pi} d\phi \int_0^{\pi} d\theta\ \sin(\theta)\,|f(k,\theta)|^2 \\ &= 4\pi\sum_{l=0}^{\infty}(2l+1)|f_l(k)|^2 = \frac{4\pi}{k^2}\sum_{l=0}^{\infty}(2l+1)\sin^2(\delta_l(k)) .\end{aligned} \tag{3.132}$$

At low momenta one can consider only the s-wave scattering ($l = 0$) and the differential cross section does not depend on the angle θ. In this case the scattering cross section is just

$$\sigma(k) = 4\pi|f_0(k)|^2 = \frac{4\pi}{k^2}\sin^2(\delta_0(k)) \,, \tag{3.133}$$

where $\delta_0(k)$ is the s-wave phase shift of the cross-section. The wave scattering length a_s, which characterizes the effective size of the target, is then defined as the following low-energy limit

$$a_s = -\lim_{k\to 0}\frac{1}{k}\tan\left(\delta_0(k)\right) . \tag{3.134}$$

In this way one then finds

$$\lim_{k\to 0}\sigma(k) = 4\pi a_s^2 \,. \tag{3.135}$$

Thus, for a very dilute and ultracold gas of atoms the cross-section $\sigma(k)$ of low-energy atoms can be indeed approximated as $\sigma(k) \simeq \sigma(0) = 4\pi a_s^2$, where the value of a_s must be obtained from the knowledge of s-wave scattering amplitude $f_0(k)$ or the s-wave phase shift $\delta_0(k)$.

All the described results crucially depend on the scattering amplitude $f(k,\theta)$. How can one explicitly calculate $f(k,\theta)$ for a given potential $V(\mathbf{r})$? Eq. (3.127) admits the formal and implicit Lippmann-Schwinger solution

$$\psi(\mathbf{r}) = e^{i\mathbf{k}\cdot\mathbf{r}} - \frac{m}{4\pi\hbar^2}\int d^3\mathbf{r}'\frac{e^{ik|\mathbf{r}-\mathbf{r}'|}}{|\mathbf{r}-\mathbf{r}'|}V(\mathbf{r}')\,\psi(\mathbf{r}') \,. \tag{3.136}$$

In far field, where

$$\frac{e^{ik|\mathbf{r}-\mathbf{r}'|}}{|\mathbf{r}-\mathbf{r}'|} \simeq \frac{e^{ikr}}{r} \,, \tag{3.137}$$

Eq. (3.136) becomes Eq. (3.128) with

$$f(k,\theta) = -\frac{m}{4\pi\hbar^2}\int d^3\mathbf{r}'\,V(\mathbf{r}')\,\psi(\mathbf{r}') \,. \tag{3.138}$$

Substituting in this formula $\psi(\mathbf{r}')$ with $e^{i\mathbf{k}\cdot\mathbf{r}'}$, we obtain the s-wave Born approximation

$$f_0(k) = -\frac{m}{4\pi\hbar^2}\int d^3\mathbf{r}'\,V(\mathbf{r}')\,e^{i\mathbf{k}\cdot\mathbf{r}'} = -\frac{m}{4\pi\hbar^2}\tilde{V}(k) \tag{3.139}$$

for the scattering amplitude, where $\tilde{V}(k)$ is the Fourier transform of $V(\mathbf{r})$. Working in the Born approximation and considering the contact potential

$$V(\mathbf{r}) = g\,\delta^{(3)}(\mathbf{r}) \tag{3.140}$$

with $\delta^{(3)}(\mathbf{r})$ the Dirac delta function in three spatial dimensions, one then finds

$$a_s = \frac{m}{4\pi\hbar^2} g \,, \tag{3.141}$$

that is an explicit relation between the s-wave scattering length a_s and the strength g of the contact interaction. Considering instead the attractive square-well potential

$$V(\mathbf{r}) = \begin{cases} -V_0 & \text{for } r < R_0 \\ 0 & \text{for } r > R_0 \end{cases} \tag{3.142}$$

it is possible to derive the expression

$$a_s = R_0 \left(1 - \frac{\tan(k_0 R_0)}{k_0 R_0} \right) \tag{3.143}$$

with $k_0 = \sqrt{mV_0/\hbar^2}$. The behavior of the scattering length a_s as a function of $k_0 R_0$ is quite remarkable: for $k_0 R_0 \to 0^+$ one has $a_s \to 0^-$, for $k_0 R_0 \to (\pi/2)^\mp$ follows that $a_S \to \pm\infty$. A divergent scattering length a_s means the formation of a bound state (also the scattering cross section diverges) and it is said that a resonance occurs. Actually, by increasing the potential depth V_0 one produces a new bound state and a resonance every time that $k_0 R_0 = (2j+1)\pi/2$ with j a natural number. This is the mechanism associated with the so-called Feshbach resonances, where the scattering length a_s diverges due to a change in the interaction potential induced by an external constant magnetic field $\mathbf{B}$. When $k_0 R_0 = j\pi$, the scattering length a_s vanishes identically, the scattering cross section is zero, and the target becomes invisible (Ramsauer-Townsend effect). Neutral alkali atoms interact through short-range van der Waals interaction $V(\mathbf{r})$ far from resonance in ultracold atomic gases, which create an environment for the exploitation of resonant scattering processes. Indeed, it is possible to adjust the effective strength of interaction by permitting particles to create a virtual bound state known as a resonance. As previously said, a Feshbach resonance allows the two-atom system to be tuned via an external magnetic field **B**. In this way the scattering length a_s can be modified from repulsive to attractive simply by changing the field **B**.

3.6.1 T-Matrix Formalism

The scattering theory can also be formulated using the transfer matrix (T-matrix) formalism, where the starting point is the Hamiltonian operator

$$\hat{H} = \hat{H}_0 + \hat{V} \ , \tag{3.144}$$

with $\hat{H}_0 = \hat{\mathbf{p}}^2/(2m_r)$, the kinetic energy operator of a particle of reduced mass $m_r = m/2$ and linear momentum $\hat{\mathbf{p}}$, while $\hat{V}$ is the interaction potential operator such that $\hat{V}|\mathbf{r}\rangle = V(\mathbf{r})|\mathbf{r}\rangle$, where $|\mathbf{r}\rangle$ is the eigenstate of the position operator $\hat{\mathbf{r}}$. We introduce $|\mathbf{k}\rangle$ as the initial state of the problem, which is assumed to be an eigenstate of the linear momentum operator $\hat{\mathbf{p}}$, i.e., $\hat{\mathbf{p}}|\mathbf{k}\rangle = \hbar\mathbf{k}|\mathbf{k}\rangle$, and it is such that

$$\hat{H}_0|\mathbf{k}\rangle = E_k|\mathbf{k}\rangle \tag{3.145}$$

with $E_k = \hbar^2k^2/(2m_r)$. Within this Dirac notation, remember that $\psi(\mathbf{r}) = \langle\mathbf{r}|\psi\rangle$, Eq. (3.127) is rewritten as

$$\left(\hat{H}_0 + \hat{V}\right)|\psi\rangle = E_k|\psi\rangle \ , \tag{3.146}$$

where $|\psi\rangle$ is the state we are looking for. The T-matrix is defined as the operator $\hat{T}$ such that

$$\hat{T}|\mathbf{k}\rangle = \hat{V}|\psi\rangle \ , \tag{3.147}$$

and its connection with the scattering amplitude $f(k,\theta)$ is given by the formula

$$f(k,\theta) = -\frac{m}{4\pi\hbar^2}T_{\mathbf{k}\mathbf{k}'} \ , \tag{3.148}$$

where

$$T_{\mathbf{k}\mathbf{k}'} = \langle\mathbf{k}|\hat{T}|\mathbf{k}'\rangle \tag{3.149}$$

is a matrix element of $\hat{T}$ with the constraint that $|\mathbf{k}| = |\mathbf{k}'| = k$. After some manipulations of the previous equations, one finds that the matrix element $T_{\mathbf{k}\mathbf{k}'}$ satisfies the T-matrix equation

$$T_{\mathbf{k}\mathbf{k}'} = V_{\mathbf{k}\mathbf{k}'} + \frac{1}{L^3}\sum_{\mathbf{q}} \frac{V_{\mathbf{k}\mathbf{q}}}{\frac{\hbar^2 k^2}{2m_r} - \frac{\hbar^2 q^2}{2m_r} + i0^+} T_{\mathbf{q}\mathbf{k}'} \ , \tag{3.150}$$

where $|\mathbf{q}\rangle$ is an intermediate state and $V_{\mathbf{k}\mathbf{k}'} = \langle \mathbf{k}|\hat{V}|\mathbf{k}'\rangle = \tilde{V}(\mathbf{k}-\mathbf{k}')$.

In the special case of contact interaction, Eq. (3.140), the T-matrix equation gives

$$T_{\mathbf{k}\mathbf{k}'} = g + g\frac{1}{L^3}\sum_{\mathbf{q}} \frac{1}{\frac{\hbar^2 k^2}{2m_r} - \frac{\hbar^2 q^2}{2m_r} + i0^+} T_{\mathbf{q}\mathbf{k}'} \ , \tag{3.151}$$

Perturbatively, at the lowest order we have

$$T_{\mathbf{k}\mathbf{k}'} = g \tag{3.152}$$

that is the Born approximation. In addition, setting $T(E_k) = T_{\mathbf{k}\mathbf{k}'}$ and working in the s-wave regime where $T_{\mathbf{k}\mathbf{k}'} \simeq T_{\mathbf{q}\mathbf{k}'}$, we obtain

$$T(E_k) = g + g\,T(E_k)\frac{1}{L^3}\sum_{\mathbf{q}} \frac{1}{E_k - \frac{\hbar^2 q^2}{2m_r} + i0^+} \ , \tag{3.153}$$

namely

$$\frac{1}{T(E_k)} = \frac{1}{g} - \frac{1}{L^3}\sum_{\mathbf{q}} \frac{1}{E_k - \frac{\hbar^2 q^2}{2m_r} + i0^+} \ . \tag{3.154}$$

Finally, identifing $T(0)$ as the renormalized interaction strength g_r, we have

$$\frac{1}{g_r} = \frac{1}{g} + \frac{1}{L^3}\sum_{\mathbf{q}} \frac{m}{\hbar^2 q^2} \ . \tag{3.155}$$

which is meaningful across a Feshbach resonance, provided the inclusion of an ultraviolet cutoff Λ for the wavenumber $q = |\mathbf{q}|$.

3.7 Interacting Bose Gas

We consider an interacting Bose gas with a contact interaction, namely

$$V(\mathbf{r}-\mathbf{r}') = g\,\delta^{(3)}(\mathbf{r}-\mathbf{r}')\;, \tag{3.156}$$

where $\delta^{(3)}(\mathbf{r})$ is the Dirac delta function in three spatial dimensions and g the strength of the interaction. By construction, from this formula it follows that

$$g = \int d^3\mathbf{r}\, V(\mathbf{r})\;. \tag{3.157}$$

Moreover, taking into account the Born formula of scattering theory for the s-wave scattering length

$$a_s = \frac{m}{4\pi\hbar^2}\int d^3\mathbf{r}\, V(\mathbf{r})\;, \tag{3.158}$$

we get

$$g = \frac{4\pi\hbar^2}{m} a_s\;. \tag{3.159}$$

The partition function $\mathcal{Z}$ of the system at temperature T can then be written as

$$\mathcal{Z} = \int \mathcal{D}[\psi^*,\psi]\;\exp\left\{-\frac{S_E[\psi^*,\psi]}{\hbar}\right\}\;, \tag{3.160}$$

where

$$S_E[\psi^*,\psi] = \int_0^{\hbar\beta} d\tau \int_{L^D} d^D\mathbf{r}\;\mathscr{L}_E(\psi^*,\psi) \tag{3.161}$$

is the Euclidean action with

$$\mathscr{L}_E = \psi^*\left[\hbar\frac{\partial}{\partial\tau} - \frac{\hbar^2}{2m}\nabla^2 + U(\mathbf{r}) - \mu\right]\psi + \frac{1}{2}\,g\,|\psi|^4\;. \tag{3.162}$$

the Euclidean Lagrangian density, and $\beta \equiv 1/(k_B T)$ with k_B being Boltzmann's constant. Notice that this Lagrangian density is equivalent to one of Eq. (3.83) because they give the same Euler-Lagrange equations. It is straightforward to verify that the Lagrangian density is invariant with respect to the global $U(1)$ transformation

$$\psi(\mathbf{r},\tau) \to \psi(\mathbf{r},\tau)\, e^{i\alpha} \tag{3.163}$$

with α space-time independent angle. The grand potential Ω of the system, which is a function of μ, T, and g, is then obtained as

$$\Omega = -\frac{1}{\beta}\ln(\mathcal{Z}) \; . \tag{3.164}$$

3.7.1 Saddle-Point Approximation

The saddle point approximation gives the partition function

$$\mathcal{Z}_0 = \exp\left\{-\frac{S_E[\psi_c^*,\psi_c]}{\hbar}\right\} , \tag{3.165}$$

where $\psi_c(\mathbf{r},\tau)$ is the "classical" field which extremizes the Euclidean action. Thus, $\psi_c(\mathbf{r},\tau)$ satisfies the Euler-Lagrange equation

$$\left(\hbar\frac{\partial}{\partial\tau} - \frac{\hbar^2}{2m}\nabla^2 + U(\mathbf{r}) - \mu + g|\psi_c(\mathbf{r},\tau)|^2\right)\psi_c(\mathbf{r},\tau) = 0 \tag{3.166}$$

The corresponding saddle-point grand potential Ω_0 of the system is clearly

$$\Omega_0 = -\frac{1}{\beta}\ln(\mathcal{Z}_0) \; . \tag{3.167}$$

Sometimes this is also called the mean-field result at zero temperature.

3.7.2 Uniform Bose Gas and Bogoliubov Spectrum

In the absence of the external potential, i.e.

$$U(\mathbf{r}) = 0 \; , \tag{3.168}$$

one finds immediately the saddle-point partition function $\mathcal{Z}_0$ of the uniform and constant bosonic field

$$\psi_c(\mathbf{r},\tau) = \psi_0 \tag{3.169}$$

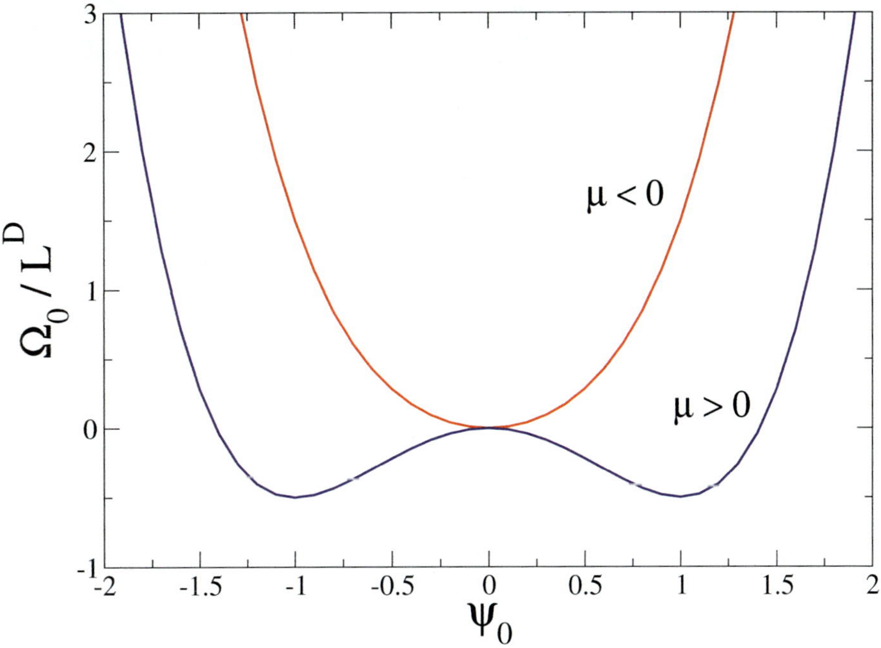

Figure 3.1
Mean-field grand potential Ω_0 as a function of the real-order parameter ψ_0 for an interacting Bose gas, see Eq. (3.172).

as

$$\mathcal{Z}_0 = \exp\left\{-\frac{S_0}{\hbar}\right\} = \exp\left\{-\beta\,\Omega_{MF}\right\} , \tag{3.170}$$

where $S_0 = S_E[|\psi_0|^2]$ and the saddle-point grand potential reads

$$\frac{\Omega_0}{L^3} = -\mu\,|\psi_0|^2 + \frac{1}{2}\,g\,|\psi_0|^4 . \tag{3.171}$$

This is the saddle-point grand potential of the uniform and interacting bosonic gas confined in a cubic box of size L.

Considering a real order parameter ψ_0, which corresponds to a breaking of the $U(1)$ symmetry, we have

$$\frac{\Omega_0}{L^3} = -\mu\,\psi_0^2 + \frac{1}{2}\,g\,\psi_0^4 . \tag{3.172}$$

The order parameter ψ_0 is then obtained by minimizing Ω_0 as

$$\left(\frac{\partial\Omega_0}{\partial\psi_0}\right)_{\mu,T,L^D} = 0 \, , \tag{3.173}$$

from which one finds the relation between order parameter and chemical potential

$$\mu = g\,\psi_0^2 \, . \tag{3.174}$$

showing that, with $g > 0$ (repulsive interaction), in the superfluid broken phase the chemical potential is positive and

$$\psi_0 = \sqrt{\frac{\mu}{g}} \, , \tag{3.175}$$

Here, we have chosen the positive ψ_0 between the two possible minima shown in Fig. 3.7.2. This corresponds to a further symmetry breaking of the $U(1)$ symmetry involving ψ_0: from a complex number to a real number to a specific real number. The breaking of the continuous $U(1)$ symmetry is called "spontaneous" because in real physical systems, the breaking of some symmetry of the Lagrangian density naturally happens at some phase transition with the help of a very small environmental noise. In our context, it is possible to prove that imposing the spontaneous symmetry breaking implies

$$\langle\psi(\mathbf{r},\tau)\rangle \neq 0 \tag{3.176}$$

and specifically

$$\langle\psi(\mathbf{r},\tau)\rangle = \psi_0 = \sqrt{\frac{\mu}{g}} \tag{3.177}$$

for a uniform 3D system of identical bosons with contact repulsive interaction below the critical temperature $T_{\rm BEC}$ of Bose-Einstein condensation. Within this framework ψ_0^2 can be identified as the number density n_0 of condensed bosons.

Inserting the relation (3.175) into Eq. (3.172), we find

$$\frac{\Omega_0}{L^3} = -\frac{\mu^2}{2\,g} \, . \tag{3.178}$$

Clearly, this equation of state is lacking important information encoded in quantum and thermal fluctuations.

Let us now consider the decomposition

$$\psi(\mathbf{r},\tau) = \psi_0 + \eta(\mathbf{r},\tau) \tag{3.179}$$

and expand the action $S[\psi,\psi^*]$ around ψ_0 up to quadratic (Gaussian) order in $\eta(\mathbf{r},\tau)$ and $\eta^*(\mathbf{r},\tau)$. This procedure corresponds, in real time, to the stationary phase approximation. We also set

$$\eta(\mathbf{r},\tau) = \sum_{n=-\infty}^{+\infty}\sum_{\mathbf{k}} \eta_{\mathbf{k},n}\,\frac{e^{i(\mathbf{k}\cdot\mathbf{r}-\omega_n\tau)}}{L^{3/2}}\,, \tag{3.180}$$

where $\mathbf{k}$ the 3-dimensional wavevector and n the quantum number of the bosonic Matsubara frequencies $\omega_n = 2\pi n/(\beta\hbar)$. Please notice that

$$\int_0^{\hbar\beta} d\tau \int_{L^3} d^3\mathbf{r}\, e^{i[(\mathbf{k}-\mathbf{k}')\cdot\mathbf{r}-(\omega_n-\omega_n')\tau]} = \hbar\beta\, L^3\, \delta_{\mathbf{k},\mathbf{k}'}\, \delta_{n,n'}\,. \tag{3.181}$$

Considering all these ingredients, we find

$$\mathcal{Z} = \mathcal{Z}_0 \int \mathcal{D}[\eta^*,\eta]\, \exp\left\{-\frac{S_g[\eta^*,\eta]}{\hbar}\right\}\,, \tag{3.182}$$

where

$$S_g[\eta^*,\eta] = \frac{\hbar}{2}\sum_{n=-\infty}^{+\infty}\sum_{\mathbf{k}} (\eta^*_{\mathbf{k},n}, \eta_{-\mathbf{k},-n})\, \hat{M}_{\mathbf{k},n} \begin{pmatrix} \eta_{\mathbf{k},n} \\ \eta^*_{-\mathbf{k},-n} \end{pmatrix} \tag{3.183}$$

is the Gaussian action of fluctuations in reciprocal space and

$$\hat{M}_{\mathbf{k},n} = \beta \begin{pmatrix} -i\hbar\omega_n + \frac{\hbar^2k^2}{2m} - \mu + 2g\psi_0^2 & g\psi_0^2 \\ g\psi_0^2 & i\hbar\omega_n + \frac{\hbar^2k^2}{2m} - \mu + 2g\psi_0^2 \end{pmatrix} \tag{3.184}$$

is the inverse fluctuation propagator. The Euclidean action can be written in a diagonal form by performing the bosonic Bogoliubov transformation

$$\begin{pmatrix} \eta_{\mathbf{k},n} \\ \eta^*_{-\mathbf{k},-n} \end{pmatrix} = \begin{pmatrix} u_{\mathbf{k}} & -v_{\mathbf{k}} \\ -v_{\mathbf{k}} & u_{\mathbf{k}} \end{pmatrix} \begin{pmatrix} b_{\mathbf{k},n} \\ b^*_{-\mathbf{k},-n} \end{pmatrix}\,, \tag{3.185}$$

where the Bogoliubov amplitudes $u_{\mathbf{k}}$ and $v_{\mathbf{k}}$ are assumed to be real. In particular, inserting the Bogoliubov transformation into the Euclidean action of Gaussian fluctuations one finds

$$\begin{aligned} S_g &= \frac{\hbar}{2}\sum_{n=-\infty}^{+\infty}\sum_{\mathbf{k}}(b^*_{\mathbf{k},n}, b_{-\mathbf{k},-n}) \\ &\times \begin{pmatrix} \beta(-i\hbar\omega_n + E_k) & 0 \\ 0 & \beta(i\hbar\omega_n + E_k) \end{pmatrix} \begin{pmatrix} b_{\mathbf{k},n} \\ b^*_{-\mathbf{k},-n} \end{pmatrix} \\ &= \hbar\sum_{\mathbf{k}}\sum_{n=-\infty}^{+\infty} \beta(-i\hbar\omega_n + E_k)|b_{\mathbf{k},n}|^2 \,. \end{aligned} \tag{3.186}$$

This result holds under the following condition

$$u_{\mathbf{k}}^2 = 1 - v_{\mathbf{k}}^2 = \frac{1}{2}\left(1 + \frac{\frac{\hbar^2 k^2}{2m} + g\psi_0^2}{E_k}\right) \tag{3.187}$$

with

$$E_k = \sqrt{\left(\frac{\hbar^2 k^2}{2m} - \mu + 2g\psi_0^2\right)^2 - g^2\psi_0^4} \tag{3.188}$$

the so-called Bogoliubov spectrum. Eq. (3.186) is a quite simple expression, similar to that one of a gas of ideal bosons. Indeed, the main idea is that the system of interacting bosonic particles is mapped into an effective system of non-interacting quasi-particles with E_k their elementary excitations. Notice that the 2×2 Bogoliubov matrix of Eq. (3.185) has the determinant equal to one, but it is not unitary. Moreover, the eigenvalues of the matrix $\hat{M}_{\mathbf{k}n}$ are not the two elements of the diagonal matrix that appears in Eq. (3.186), however the determinant of the two matrices is the same. Actually, within the path integral approach, one could avoid the use of the Bogoliubov transformation by calculating directly the determinant of the matrix $\hat{M}_{\mathbf{k}n}$.

By using $\psi_0 = \sqrt{\mu/g}$ the Bogoliubov spectrum (3.188) becomes gapless, namely

$$E_k = \sqrt{\frac{\hbar^2 k^2}{2m}\left(\frac{\hbar^2 k^2}{2m} + 2\mu\right)} \,. \tag{3.189}$$

Remarkably, one also gets

$$E_k \simeq c_B \, \hbar k \qquad \text{for small } k \tag{3.190}$$

with $c_B = \sqrt{\mu/m}$ is the Bogliubov sound velocity.

Integrating over the Bogoliubov complex fields $b_{\mathbf{k},n}$ and $b^*_{\mathbf{k},n}$ one easily derives the Gaussian partition function

$$\mathcal{Z}_g = \frac{1}{\prod_{\mathbf{k}} \prod_{n=-\infty}^{+\infty} \beta(-i\hbar\omega_n + E_k)} = \frac{1}{\prod_{\mathbf{k}}(1 - e^{-\beta E_k})} \tag{3.191}$$

and the Gaussian grand potential

$$\Omega_g = \frac{1}{\beta} \sum_{\mathbf{k}} \sum_{n=-\infty}^{+\infty} \ln\left[\beta(-i\hbar\omega_n + E_k)\right] = \frac{1}{\beta} \sum_{\mathbf{k}} \ln\left(1 - e^{-\beta E_k}\right) . \tag{3.192}$$

Thus, the mean-field grand potential of our system of interacting bosons reads

$$\Omega_{MF} = \Omega_0 + \Omega_g = -\frac{\mu^2}{2\, g} L^3 + \sum_{\mathbf{k}} \ln\left(1 - e^{-\beta E_k}\right) . \tag{3.193}$$

3.7.3 Zero-Point Energy of Bogoliubov Excitations

In our treatment of the bosonic field, we have not considered the zero-point energy of bosonic collective excitations and, specifically, the zero-temperature contribution of quantum Gaussian fluctuations associated with the Bogoliubov elementary excitations. This contribution reads

$$\Omega_g^{(0)} = \frac{1}{2} \sum_{\mathbf{k}} (E_k - \frac{\hbar^2 k^2}{2m} - \mu) \tag{3.194}$$

with E_k given by Eq. (3.189). This term originates from the Gaussian action when one interchanges the order of the Bogoliubov complex fields $b_{\mathbf{k},n}$ and $b^*_{\mathbf{k},n}$. One must do it in the space-time discretized version of the path integral formulation, performing the continuum time limit only after the interchange of the fields. In this way a convergence factor $e^{i\omega_n 0^+}$ comes out and determines the appearance of the zero-point energy of the interacting field. Notice that for the non-interacting bosonic field in the BEC phase one has $\mu = 0$, $E_k = \hbar^2 k^2/(2m)$, and consequently $\Omega_g^{(0)} = 0$. For the sake of completeness, we observe that one can deduce the zero-point energy starting from the continuum version of the path integral and using an alternative treatment (without convergence factors) of the infinite product of bosonic Matsubara frequencies.

3.7.4 Dimensional Regularization

In the continuum $\sum_{\mathbf{k}} \to L^3 \int d^3\mathbf{k}/(2\pi)$ Eq. (3.194) is ultraviolet divergent. We use dimensional regularization to extract the finite contribution of $\Omega_g^{(0)}$. The Hooft-Veltman conjecture states that integrals over a polynomial identically vanish, i.e.

$$\int_0^{+\infty} dk\, k^{D-1} \, (k^2)^{n-1} \, , \qquad n = 0, 1, 2, ... \tag{3.195}$$

with D belonging to the set $\mathbb{C}$ of complex numbers. Thus, the integral of the terms $-\hbar^2 k^2/(2m) - \mu$ is regularized to zero and we can write

$$\Omega_g^{(0)} = \frac{1}{2} \sum_{\mathbf{k}} E_k \, . \tag{3.196}$$

For this remaining divergent part, working in D spatial dimensions we get

$$\frac{\Omega_g^{(0)}}{L^D} = \frac{1}{2} \frac{S_D}{(2\pi)^D} \int_0^{+\infty} dk \; k^{D-1} \sqrt{\frac{\hbar^2 k^2}{2m} \left(\frac{\hbar^2 k^2}{2m} + 2\mu \right)} \tag{3.197}$$

with $S_D = 2\pi^{D/2}/\Gamma(D/2)$ is solid angle in D dimensions and $\Gamma(x)$ the Euler gamma function. As expected, the integral is ultraviolet divergent at any integer dimension D. We rewrite Eq. (3.197) for the zero-point energy of a repulsive Bose gas in dimension D as

$$\frac{\Omega_g^{(0)}}{L^D} = \frac{S_D (2\mu)^{\frac{D}{2}+1}}{4(2\pi)^D} \left(\frac{2m}{\hbar^2} \right)^{\frac{D}{2}} B\Big(-\frac{D+2}{2}, \frac{D+1}{2} \Big) \, , \tag{3.198}$$

where $B(x,y)$ is the Euler beta function

$$B(x,y) = \int_0^{+\infty} dt \frac{t^{x-1}}{(1+t)^{x+y}}, \qquad Re(x), Re(y) > 0 \qquad (3.199)$$

which may be analytically continued to complex values of x and y as

$$B(x,y) = \frac{\Gamma(x)\,\Gamma(y)}{\Gamma(x+y)} \qquad (3.200)$$

with $\Gamma(z)$ the Euler gamma function. This is the crucial step, where one assumes that D must be interpreted as a complex number. Rewriting Eq. (3.197) by using Eq. (3.200), we obtain

$$\frac{\Omega_g^{(0)}}{L^D} = \frac{S_D(2\mu)^{\frac{D}{2}+1}}{4(2\pi)^D} \left(\frac{2m}{\hbar^2}\right)^{\frac{D}{2}} \frac{\Gamma(-\frac{D+2}{2})\,\Gamma(\frac{D+1}{2})}{\Gamma(-\frac{1}{2})} . \qquad (3.201)$$

This expression is now finite setting $D = 1$ or $D = 3$, while it is still divergent if $D = 2$ since $\Gamma(p)$ diverges for integers $p \leq 0$. In particular, setting $D = 3$ in Eq. (3.201) we get

$$\frac{\Omega_g^{(0)}}{L^3} = \frac{8}{15\pi^2}(\frac{m}{\hbar^2})^{3/2}\mu^{5/2} . \qquad (3.202)$$

The total grand potential of the three-dimensional Bose gas, including the zero-point energy of Gaussian fluctuations, is given by

$$\frac{\Omega}{L^3} = -\frac{\mu^2}{2\,g} + \frac{8}{15\pi^2}(\frac{m}{\hbar^2})^{3/2}\mu^{5/2} + \frac{1}{\beta L^3}\sum_{\mathbf{k}} \ln\left(1 - e^{-\beta E_k}\right) . \quad (3.203)$$

This result was obtained for the first time in 1957 by Tsung-Dao Lee, Kerson Huang, and Chen Ning Yang without invoking dimensional regularization. Usually, the grand potential without the zero-point energy term is called the mean-field grand potential. In other words, the mean field includes the temperature-independent saddle-point contribution, but also the thermal Gaussian fluctuations. From this point of view, the Lee-Hang-Yang term is a beyond-mean-field correction.

3.7.5 Entropy, Pressure and Number Density

Having the grand potential Ω of Eq. (3.203) many other thermodynamical quantities follow from the relation

$$d\Omega = -S\,dT - P\,dV - N\,d\mu \ , \tag{3.204}$$

where $V = L^3$ is the volume. In particular, the entropy is

$$S = -\left(\frac{\partial \Omega}{\partial T}\right)_{\mu,V} \ . \tag{3.205}$$

The pressure is

$$P = -\left(\frac{\partial \Omega}{\partial V}\right)_{\mu,T} = -\frac{\Omega}{V} \ , \tag{3.206}$$

where the second equality follows from the extensiveness of the energy (ensured for a homogeneous system by its additivity). Finally the number density of particles reads

$$n = -\frac{1}{V}\left(\frac{\partial \Omega}{\partial \mu}\right)_{T,V} \ . \tag{3.207}$$

From Eqs. (3.203) and (3.207) we deduce

$$n = \frac{\mu}{g} - \frac{4}{3\pi^2}\left(\frac{m}{\hbar^2}\right)^{3/2}\mu^{3/2} - \frac{1}{L^3}\sum_{\mathbf{k}} \frac{\frac{\partial E_k}{\partial \mu}}{e^{\beta E_k} - 1} \ . \tag{3.208}$$

This is an implicit formula for the equation of state, namely the chemical potential μ as a function of the total number density n and the temperature T, because E_k is given by Eq. (3.189).

3.7.6 Condensate Fraction

The condensate number density is defined as

$$n_0 = \psi_0^2 \tag{3.209}$$

and consequently $\mu = g\psi_0^2 = gn_0$. For computing the condensate fraction n_0/n from the knowledge of the grand canonical potential Ω it is, however, necessary to consider the chemical potential μ

and the condensate density $n_0 = \psi_0^2$ as independent variables. In other words, one must begin from the Bogoliubov spectrum given by Eq. (3.188), which can be rewritten as

$$E_k = \sqrt{\left(\frac{\hbar^2 k^2}{2m} - \mu + 2g\psi_0^2\right)^2 - g^2 n_0^2} \,. \tag{3.210}$$

Let us start with the zero-temperature case $T = 0$. We have $\Omega_g^{(T)} = 0$ and $\Omega = \Omega_0 + \Omega_g^{(0)}$, therefore

$$\begin{aligned} n(n_0, T=0) &= -\frac{1}{V}\frac{\partial \Omega(\mu, n_0, T=0)}{\partial \mu}\bigg|_{\mu = g n_0} \\ &= n_0 + n_g^{(0)}(n_0) \end{aligned} \tag{3.211}$$

Note that the substitution $\mu = g n_0$ occurs after having calculated the derivative. We have

$$\begin{aligned} n_g^{(0)}(n_0) &= -\frac{1}{V}\frac{\partial \Omega_g^{(0)}(\mu, n_0)}{\partial \mu}\bigg|_{\mu = g n_0} \\ &= -\frac{1}{2V}\sum_{\mathbf{q}} \frac{\partial E_k(\mu, n_0)}{\partial \mu}\bigg|_{\mu = g n_0} \\ &= \frac{1}{2}\frac{4\pi}{(2\pi)^3}\int_0^\infty dk\, k^2 \frac{\frac{\hbar^2 k^2}{2m} + g n_0}{E_k(n_0)} \end{aligned} \tag{3.212}$$

with $E_k(n_0)$ given by (3.188). After the regularization of the ultraviolet divergence, we obtain

$$n_g^{(0)} = \frac{1}{3\pi^2}\left(\frac{m}{\hbar^2} g n_0\right)^{3/2} \tag{3.213}$$

from which it follows

$$n = n_0\left[1 + \frac{8}{3\sqrt{\pi}}\left(a_s n_0^{1/3}\right)^{3/2}\right] \tag{3.214}$$

and for the condensate fraction

$$\frac{8}{3\sqrt{\pi}}(a_s n^{1/3})^{3/2}\left(\frac{n_0}{n}\right)^{3/2} + \frac{n_0}{n} = 1 \tag{3.215}$$

where $a_s n^{1/3}$ is the gas parameter. Solving this equation numerically shows that in the interacting case, the condensate fraction at $T = 0$ is less than one (quantum depletion) and is a decreasing function of gas parameter. For small values of the gas parameter, we can approximate $(n_0/n)^{3/2} \simeq 1$ in the previous equation, finding

$$\frac{n_0}{n} \simeq 1 - \frac{8}{3\sqrt{\pi}} \left(a_s n^{1/3}\right)^{3/2} , \tag{3.216}$$

that is the formula obtained for the first time by Bogoliubov in 1947.

At finite temperature, i.e., with $T > 0$, the grand potential is $\Omega = \Omega_0 + \Omega_g^{(0)} + \Omega_g^{(T)}$, and consequently

$$\begin{aligned} n(n_0, T) &= -\frac{1}{V} \frac{\partial \Omega(\mu, n_0, T)}{\partial \mu}\bigg|_{\mu = g n_0} \\ &= n_0 + n_g^{(0)}(n_0) + n_g^{(T)}(n_0, T) , \end{aligned} \tag{3.217}$$

where

$$\begin{aligned} n_g^{(T)}(n_0, T) &= -\frac{1}{V} \frac{\partial \Omega_g^{(T)}(\mu, n_0, T)}{\partial \mu}\bigg|_{\mu = g n_0} \\ &= -\frac{1}{\beta V} \sum_{\mathbf{q}} \frac{\partial}{\partial \mu} \ln\left(1 - \mathrm{e}^{-\beta E_k(n_0)}\right)\bigg|_{\mu = g n_0} \\ &= \frac{4\pi}{(2\pi)^3} \int_0^\infty dk\, k^2 \frac{\frac{\hbar^2 k^2}{2m} + g n_0}{\left[\mathrm{e}^{\beta E_k(n_0)} - 1\right] E_k(n_0)} \end{aligned} \tag{3.218}$$

with

$$E_k(n_0) = \sqrt{\frac{\hbar^2 k^2}{2m} \left(\frac{\hbar^2 k^2}{2m} + 2 g n_0\right)} . \tag{3.219}$$

The equation for the total number density n as a function of the condensate density n_0 and T is then

$$\begin{aligned} n(n_0, T) &= n_0 \left[1 + \frac{8}{3\sqrt{\pi}} \left(a_s n_0^{1/3}\right)^{3/2}\right] \\ &+ \frac{1}{2\pi^2} \int_0^\infty dk\, k^2 \frac{\frac{\hbar^2 k^2}{2m} + g n_0}{\left[\mathrm{e}^{\beta E_k(n_0)} - 1\right] E_k(n_0)} . \end{aligned} \tag{3.220}$$

To obtain an expression for the condensate fraction, it is necessary to numerically evaluate the integral on the right side, working iteratively until obtaining the desired value for the total density. However, in the weak-coupling regime and low-temperature regime, where for the thermal part, one can safely write

$$E_k(n_0) \simeq \frac{\hbar^2 k^2}{2m} + g n_0 \,, \tag{3.221}$$

from Eqs. (3.216) and (3.220) one deduces the very simple analytical formula

$$\frac{n_0}{n} \simeq 1 - \frac{8}{3\sqrt{\pi}} \left(a_s n^{1/3}\right)^{3/2} - \frac{1}{2\pi^2} \int_0^\infty dk\, k^2 \frac{1}{e^{\beta E_k(n)} - 1} \,. \tag{3.222}$$

Note that using Eq. (3.220) with $n_0 = 0$ and $T = T_{\mathrm{BEC}}$ one finds that T_{BEC} is one of the non-interacting bosons. This simply means that to get some dependence on the interaction strength g in the critical temperature T_{BEC} of Bose-Einstein condensation for a uniform gas of identical bosons one needs to include beyond Gaussian corrections.

3.8 Superfluidity

Superfluidity was discovered in 1937 by Pyotr Kapitza, John Allen, and Don Misener, who found that, at atmospheric pressure, below a critical temperature $T_s = 2.16$ Kelvin helium 4 (^{4}He) not only remains liquid but it also shows zero viscosity. In particular, the experiments with ^{4}He show that at the critical temperature T_s the specific heat diverges. In 1938 Fritz London gave first theoretical explanation of the superfluidity of helium 4 on the basis of Bose-Einstein condensation (BEC). However, Lev Landau in 1941 was able to describe the superfluidity without using the BEC explicitly. Within the two-fluid model of Tisza-Landau, below T_λ, Helium 4 is characterized by an inviscid superfluid component and a viscous normal component.

3.8.1 Landau Criterion

According to Landau the superfluidity of a bosonic system is a consequence of the peculiar dispersion relation of the quasi-particles. Indeeed, a superfluid is characterized by quasi-particles with a dispersion relation E_k that is not that one of free particles, i.e., $E_k = \hbar^2 k^2/(2m)$. For weakly-interacting bosons, we have found the Bogoliubov dispersion relation, Eq. (3.189), which becomes the phonon spectrum $E_k = c_B \hbar k$ for a small linear momentum $\hbar k$, with c_B the Bogoliubov speed of sound. In the case of superfluid ^{4}He the dispersion relation E_k is more complicated, but at low momenta it has a phonon-like behavior.

Let us consider a macroscopic particle of mass M that is moving with velocity $\mathbf{v}$ inside the superfluid. In the scattering between the macroscopic particle and one quasi-particle of the superfluid, the laws of conservation of linear momentum and energy

$$\begin{aligned} \frac{1}{2} M v^2 &= \frac{1}{2} M (v')^2 + E_k \ , \quad (3.223) \\ M\mathbf{v} &= M\mathbf{v}' + \hbar\mathbf{k} \ , \quad (3.224) \end{aligned}$$

where, before the collision, the energy and momentum of the quasi-particle are zero. Combining the two equations, one finds

$$\mathbf{v} \cdot \hbar\mathbf{k} = E_k + \frac{\hbar^2 k^2}{2M} \ . \quad (3.225)$$

If the mass M is very large, one gets

$$v = \frac{E_k}{k \ \cos(\alpha)} \ , \quad (3.226)$$

where α is the angle between $\mathbf{k}$ and $\mathbf{v}$. The equality (3.226) holds only if v is larger than the minimum of $E_k/(\hbar k)$ for any choice of $\mathbf{k}$. The Landau criterion says that at zero temperature the superfluid is lost only if the velocity v of the macroscopic particle is larger than the critical velocity

$$v_{\text{crit}} = min_{\mathbf{k}} \left(\frac{E_k}{\hbar k} \right) \ . \quad (3.227)$$

Clearly, in the case of the Bogoliubov spectrum one finds $v_{\text{crit}} = c_s$, while for the free-particle spectrum, one gets $v_{\text{crit}} = 0$.

3.8.2 Superfluid Fraction

The quasi-particles with energy spectrum E_k play a crucial role in the two-fluid model of Landau, inspired by similar models of Lazlo Tisza for superfluids, and Fritz and Heinz London for superconductors. According to the two-fluid model, at thermal equilibrium the superfluid system is characterized by the total number density

$$n = n_s + n_n \ , \tag{3.228}$$

where n_s is the superfluid number density and n_n is the normal number density.

At temperature T, in the rest frame of the superfluid, the normal current density of mass is given by

$$\mathbf{j}_n = m n_n \mathbf{v} = \int \frac{d^3\mathbf{k}}{(2\pi)^3} \, \hbar\mathbf{k} \, f_B(E_k - \hbar\mathbf{k}\cdot\mathbf{v}) \ , \tag{3.229}$$

where $\mathbf{v}$ is the velocity of the normal fluid and $f_B(E) = \frac{1}{e^{E/(k_B T)}-1}$ is the Bose distribution.

Assuming a small velocity $\mathbf{v}$ and Taylor-expanding the previous formula with respect to $\mathbf{v}$ to the first order, one finds

$$n_n(T) = -\frac{1}{3} \int \frac{d^3\mathbf{k}}{(2\pi)^3} \frac{\hbar^2 k^2}{m} \frac{d}{dE} \left(\frac{1}{e^{\frac{E_k}{k_B T}} - 1} \right) \ , \tag{3.230}$$

taking into account that $\int d^D\mathbf{k}\, \mathbf{k}(\mathbf{k}\cdot\mathbf{v})F(p) = (1/D)(\int d^D\mathbf{k}\, k^2 F_k)\mathbf{v}$. Thus, the thermal activation of quasi-particles increases the normal component of the superfluid. Clearly, the superfluid density then reads

$$n_s(T) = n - n_n(T) = n - \frac{1}{3k_B T} \frac{1}{2\pi^2} \int_0^{+\infty} dk k^2 \, \frac{\hbar^2 k^2}{m} \frac{e^{\frac{E_k}{k_B T}}}{\left(e^{\frac{E_k}{k_B T}} - 1\right)^2} \tag{3.231}$$

and the critical temperature T_s of the superfluid-normal phase transition is obtained setting $n_s(T_s) = 0$.

In three spatial dimensions, experimentally, one finds for ^{4}He that the critical temperature T_s of the superfluid-to-normal phase transition coincides the the critical temperature $T_{\rm BEC}$ of Bose-Einstein condensation. However, in two spatial dimensions, quite generally for bosonic systems $T_s \neq T_{\rm BEC} = 0$. Moreover, in one spatial dimension, for interacting bosons $T_s \neq 0$ while $T_{\rm BEC}$ of Bose-Einstein condensation does not exist. In three spatial dimensions, at low temperatures, the spectrum E_k can be safely approximated as phononic, i.e., $E_k = c_B \hbar q$, and the superfluid fraction reads

$$\frac{n_s}{n} = 1 - \frac{2\pi^2}{45\hbar^3 n m c_B^5} (k_B T)^4 \, . \tag{3.232}$$

At zero temperature the superfluid fraction of an interacting bosonic system is equal to one, while the condensed fraction is smaller than one. For instance, experimental data and numerical simulations suggest that for Helium 4 the condensate fraction is below 10% close to zero temperature.

3.9 Gross-Pitaevskii Equation

Let us consider again a system of bosons with a repulsive contact interaction, i.e., let us set $g > 0$ in the Euclidean Lagrangian density

$$\mathscr{L}_E = \psi^* \left[\hbar \frac{\partial}{\partial \tau} - \frac{\hbar^2}{2m} \nabla^2 - \mu + U(\mathbf{r}) \right] \psi + \frac{1}{2} g \, |\psi|^4 \, , \tag{3.233}$$

where $U(\mathbf{r})$ is the external trapping potential.

In this inhomogeneous case the partition function of the mean-field (saddle-point) order parameter $\psi_c(\mathbf{r}, \tau)$ reads

$$\mathcal{Z}_0 = \exp\left\{ -\frac{S_c}{\hbar} \right\} = \exp\left\{ -\beta \, \Omega_{MF} \right\} \, , \tag{3.234}$$

where

$$S_c = S_E[\psi_c^*(\mathbf{r}, \tau), \psi_c(\mathbf{r}, \tau)] = \int_0^{\hbar\beta} d\tau \int d^3\mathbf{r} \; \mathscr{L}_E(\psi_c, \psi_c^*) \tag{3.235}$$

and $\psi_c(\mathbf{r},\tau)$ is the "classical" field which minimizes the action functional, namely such that

$$\delta S_E[\psi_c^*(\mathbf{r},\tau),\psi_c(\mathbf{r},\tau)] = 0 \; . \tag{3.236}$$

The minimization of the Euclidean action S_E gives the Euler-Lagrange equation

$$\hbar\frac{\partial}{\partial\tau}\psi_c(\mathbf{r},\tau) = \left[-\frac{\hbar^2}{2m}\nabla^2 - \mu + U(\mathbf{r}) + g\,|\psi_c(\mathbf{r},\tau)|^2\right]\psi_c(\mathbf{r},\tau) \; . \tag{3.237}$$

This is the Gross-Pitaevskii equation with imaginary time τ. Moreover, at the mean-field level, the number of bosons reads

$$N = -\frac{\partial\Omega_{MF}}{\partial\mu} = \int d^3\mathbf{r}\,|\psi_c(\mathbf{r},\tau)|^2 \; . \tag{3.238}$$

Going back to real time $t = -i\tau$, we obtain

$$i\hbar\frac{\partial}{\partial t}\psi_c(\mathbf{r},t) = \left[-\frac{\hbar^2}{2m}\nabla^2 - \mu + U(\mathbf{r}) + g\,|\psi_c(\mathbf{r},t)|^2\right]\psi_c(\mathbf{r},t) \; , \tag{3.239}$$

that is the real-time (time-dependent) Gross-Pitaevskii equation, with the normalization condition

$$N = \int d^3\mathbf{r}\,|\psi_c(\mathbf{r},t)|^2 \; . \tag{3.240}$$

Immediately one deduces the corresponding stationary Gross-Pitaevskii equation

$$\left[-\frac{\hbar^2}{2m}\nabla^2 + U(\mathbf{r}) + g\,|\psi_0(\mathbf{r})|^2\right]\psi_0(\mathbf{r}) = \mu\,\psi_0(\mathbf{r}) \tag{3.241}$$

where we have used the symbol ψ_0 instead of ψ_c to stress the fact that the "classical" Schrödinger field is time independent.

Quite remarkably, it is possible to find analytical solutions of this nonlinear Schrödinger equation. As an example, we consider a Bose-Einstein condensate made of N ultracold atoms with mass m and scattering length a_s, under external harmonic confinement $U(r) = m\omega^2 r^2/2$ of frequency ω. Neglecting the kinetic term in the Gross-Pitaevskii equation, we immediately obtain the density profile $n_0(\mathbf{r}) = |\psi_0(\mathbf{r})|^2$ and the chemical potential μ of the system. In fact, neglecting the kinetic term (second derivatives) in the Gross-Pitaevskii equation, one finds the density profile

$$n_0(\mathbf{r}) = |\psi_0(\mathbf{r})|^2 = \frac{1}{g}\left(\mu - \frac{m}{2}\omega^2 r^2\right)\Theta\left(\mu - \frac{m}{2}\omega^2 r^2\right) \tag{3.242}$$

with $\Theta(x)$ the Heaviside step function. Imposing the condition

$$N = \int d^3\mathbf{r}\, n_0(\mathbf{r}) \tag{3.243}$$

which fixes the chemical potential μ, after integration, one gets

$$\mu = \frac{1}{2}\hbar\omega\left(\frac{15Na_s}{a_H}\right)^{2/5} \tag{3.244}$$

with $a_H = \sqrt{\hbar/(m\omega)}$ the characteristic length of the isotropic harmonic potential. Notice that we have used the formula $g = 4\pi\hbar^2 a_s/m$ which relates the interaction strength g to the s-wave scattering length.

It is relevant to emphasize that the Bogoliubov spectrum can be deduced also from the time-dependent Gross-Pitaevskii equation. Let us insert the ansatz

$$\psi_c(\mathbf{r}, t) = \psi_0(\mathbf{r}) + \eta(\mathbf{r}, t)\ . \tag{3.245}$$

into the Eq. (3.239), taking into account Eq. (3.241). Under the assumption that $\eta(\mathbf{r}, t)$ is a small perturbation around the stationary and real solution $\psi_0(\mathbf{r})$, i.e.

$$|\psi|^2\psi \simeq \psi_0^3 + 2\psi_0^2\,\eta + \psi_0^2\,\eta^* \tag{3.246}$$

neglecting quadratic and cubic terms with respect to η, we obtain a differential equation for the field $\eta(\mathbf{r}, t)$. In particular

$$\begin{aligned} i\hbar\frac{\partial}{\partial t}\eta(\mathbf{r}, t) &= \left[-\frac{\hbar^2}{2m}\nabla^2 - \mu + U(\mathbf{r}) + 2g\,\psi_0(\mathbf{r})^2\right]\eta(\mathbf{r}, t) \\ &\quad + g\psi_0(\mathbf{r})^2\,\eta^*(\mathbf{r}, t)\ . \end{aligned} \tag{3.247}$$

In the absence of the external potential, i.e., $U(\mathbf{r}) = 0$, we get $\mu = g\psi_0^2$ and also

$$i\hbar\frac{\partial}{\partial t}\eta(\mathbf{r},t) = \left[-\frac{\hbar^2}{2m}\nabla^2 + U(\mathbf{r}) + g\,\psi_0^2\right]\eta(\mathbf{r},t) + g\psi_0^2\,\eta^*(\mathbf{r},t)\ . \tag{3.248}$$

The fluctuation field can be written as

$$\eta(\mathbf{r},t) = \sum_{\mathbf{k}}\left[u_k\,e^{i(\mathbf{k}\cdot\mathbf{r}-\omega_k t)} + v_k\,e^{-i(\mathbf{k}\cdot\mathbf{r}-\omega_k t)}\right]\ , \tag{3.249}$$

By inserting Eq. (3.249) into Eq. (3.248), we obtain the Bogoliubov-de Gennes algebric equations

$$L_k\,u_k + g\psi_0^2\,v_k = \hbar\omega_k\,u_k\ , \tag{3.250}$$

$$L_k\,v_k + g\psi_0^2\,u_k = -\hbar\omega_k\,v_k\ , \tag{3.251}$$

where

$$L_k = \frac{\hbar^2 k^2}{2m} + g\psi_0^2\ . \tag{3.252}$$

It is then straightforward to find that

$$E_k = \hbar\omega_k = \sqrt{\frac{\hbar^2 k^2}{2m}\left(\frac{\hbar^2 k^2}{2m} + 2g\psi_0^2\right)} \tag{3.253}$$

that is indeed the Bogoliubov spectrum.

3.9.1 Superfluid Hydrodynamics

The zero-temperature time-dependent Gross-Pitaevskii equation (3.239) can be rewritten as the equations of superfluid hydrodynamics. In fact, setting

$$\psi_c(\mathbf{r},t) = n_c(\mathbf{r},t)^{1/2}\,e^{i\theta_c(\mathbf{r},t)}\ , \tag{3.254}$$

and inserting this formula into Eq. (3.239) one finds

$$\frac{\partial}{\partial t}n_c + \boldsymbol{\nabla}\cdot(n_c\,\mathbf{v}_c) = 0\ , \tag{3.255}$$

$$m\frac{\partial}{\partial t}\mathbf{v}_c + \boldsymbol{\nabla}\left[\frac{1}{2}mv_c^2 + U(\mathbf{r}) + gn_c - \frac{\hbar^2}{2m}\frac{\nabla^2\sqrt{n}}{\sqrt{n_c}}\right] = 0\ , \tag{3.256}$$

where

$$\mathbf{v}_c(\mathbf{r},t) = \frac{\hbar}{m}\boldsymbol{\nabla}\theta_c(\mathbf{r},t) \tag{3.257}$$

is the local velocity field, that is (by definition) irrotational, i.e., such that

$$\boldsymbol{\nabla} \wedge \mathbf{v}_c = \mathbf{0} \ . \tag{3.258}$$

Eqs. (3.255) and (3.256) are, respectively, the equation of continuity and the equation of conservation of linear momentum for an irrotational and inviscid fluid. The zero-temperature equation of the state of this superfluid, i.e., the local chemical potential as a function of the local density and its derivatives can be written as

$$\mu(n_c, \nabla^2 n_c) = g n_c - \frac{\hbar^2}{2m}\frac{\nabla^2\sqrt{n_c}}{\sqrt{n_c}} \ , \tag{3.259}$$

where the second term, which is usually called quantum pressure, becomes negligible in the high-density regime. In 1941 Lev Landau, on the basis of previous important theoretical hints of Fritz London and Laszlo Tisza, developed the two-fluid theory of superfluid ^{4}He. At zero temperature this two-fluid theory gives Eqs. (3.255) and (3.256). We observe that quantum effects are encoded not only in this equation of state but also into the properties of the local field $\mathbf{v}_c(\mathbf{r},t)$: it is proportional to the gradient of a scalar field, $\theta_c(\mathbf{r},t)$, that is the angle of the phase of the single-valued complex wavefunction $\psi_c(\mathbf{r},t)$. Consequently, $\mathbf{v}_c(\mathbf{r},t)$ satisfies the equation

$$\oint_C \mathbf{v}_c \cdot d\mathbf{r} = 2\pi\frac{\hbar}{m}k \tag{3.260}$$

for any closed contour C, with k an integer number. In other words, the circulation is quantized in units of $\hbar/m$, and this property is strictly related to the existence of quantized vortices. This result is due to the fact the the comples wavefuction $\psi_c(\mathbf{r},t)$ of Eq. (3.254) is single valued and $\theta_c(\mathbf{r},t)$ is an angle. Consequently, we have

$$\oint_C \mathbf{v}_c \cdot d\mathbf{r} = \frac{\hbar}{m}\oint_C \boldsymbol{\nabla}\theta_c \cdot d\mathbf{r} = \frac{\hbar}{m}\oint_C d\theta_c = \frac{\hbar}{m}2\pi k \ . \tag{3.261}$$

For the sake of completeness, we observe that, in general, for a rotational and viscous fluid with quite generic zero-temperature equation of state $\mu(n, \nabla n, \nabla^2 n, ...)$ the equations of hydrodynamics are given by

$$
\begin{aligned}
\frac{\partial}{\partial t} n + \boldsymbol{\nabla} \cdot (n\,\mathbf{v}) &= 0 \qquad (3.262) \\
m\frac{\partial}{\partial t}\mathbf{v} - \eta\nabla^2\mathbf{v} & \\
+\boldsymbol{\nabla}\left[\frac{1}{2}mv^2 + U(\mathbf{r}) + \mu(n, \nabla n, \nabla^2 n, ...)\right] &= m\mathbf{v}\wedge(\boldsymbol{\nabla}\wedge\mathbf{v}) \qquad (3.263)
\end{aligned}
$$

where η is the viscosity and a rotational term appears on the right side of Eq. (3.263). These equations are called zero-temperature Navier-Stokes equations. Any generic fluid (in the collisional regime) satisfies these equations. Also, a superfluid is described by Eqs. (3.262) and (3.263), but only if $\eta = 0$ and under the additional constraints given by Eqs. (3.258) and (3.260).

Let us consider Eqs. (3.255) and (3.256) assuming that $U(\mathbf{r}) = 0$. We set

$$
\begin{aligned}
n_c(\mathbf{r}, t) &= n_0 + \delta n_c(\mathbf{r}, t)\,, \qquad (3.264) \\
\mathbf{v}_c(\mathbf{r}, t) &= \mathbf{0} + \delta\mathbf{v}_c(\mathbf{r}, t)\,, \qquad (3.265)
\end{aligned}
$$

where $\delta n_c(\mathbf{r}, t)$ and $\delta\mathbf{v}_c(\mathbf{r}, t)$ represent small variations with respect to the uniform and constant stationary configuration n_0. In this way, neglecting quadratic terms in the variations (linearization) from Eqs. (3.255) and (3.256), we get the linear equations of motion

$$
\begin{aligned}
\frac{\partial}{\partial t}\delta n_c + n_{eq}\boldsymbol{\nabla}\cdot\delta\mathbf{v}_c &= 0\,, \qquad (3.266) \\
\frac{\partial}{\partial t}\delta\mathbf{v}_c + \frac{c_B^2}{n_0}\boldsymbol{\nabla}\delta n - \frac{\hbar^2}{4m^2 n_0}\boldsymbol{\nabla}(\nabla^2\delta n_c) &= \mathbf{0}\,, \qquad (3.267)
\end{aligned}
$$

where c_B is the sound velocity of the bosonic superfluid, given by

$$
mc_B^2 = g\,n_0\,. \qquad (3.268)
$$

The linear equations of motion can be arranged in the form of the following wave equation

$$
\left[\frac{\partial^2}{\partial t^2} - c_B^2\nabla^2 + \frac{\hbar^2}{4m^2}\nabla^4\right]\delta n_c(\mathbf{r}, t) = 0\,. \qquad (3.269)
$$

This wave equation admits monochromatic plane-wave solutions, where the frequency ω and the wave vector $\mathbf{k}$ are related by the dispersion formula $\omega = \omega_k$ given by

$$E_k = \hbar\,\omega_k = \sqrt{\frac{\hbar^2 k^2}{2m}\left(\frac{\hbar^2 k^2}{2m} + 2mc_B^2\right)} \ . \tag{3.270}$$

We have thus obtained, in a different third way, the Bogoliubov spectrum of elementary excitations.

3.9.2 Quantized Vortices

We have seen that the quantization of circulation given by Eq. (3.260) can be explained taking into account that the field $\psi_c(\mathbf{r},t)$ of the Gross-Pitaevskii equation is complex and its phase $\theta_c(\mathbf{r},t)$ is related to the velocity field $\mathbf{v}_c(\mathbf{r},t)$ according to Eq. (3.257). In full generality we have

$$\oint_{\mathcal{C}} d\theta_c = \oint_{\mathcal{C}} \boldsymbol{\nabla}\theta_c(\mathbf{r},t)\cdot d\mathbf{r} = 2\pi \sum_j q_j \ , \tag{3.271}$$

with q_j the integer quantum number, also called topological charge, of the j-th vortex. Here, the closed contour $\mathcal{C}$ encircles a finite number of quantized vortices, each of them characterized by a point where the phase $\theta_c(\mathbf{r},t)$ is singular.

Let us consider the simplest example: the single vortex line. For a three-dimensional Bose-Einstein condensate wavefunction, we have

$$\psi_c(x,y,z) = |\psi_c(x,y,z)|\ e^{i\theta(x,y,z)} \ , \tag{3.272}$$

where

$$n_c(x,y,z) = |\psi_c(x,y,z)|^2 \ , \qquad \mathbf{v}_c(x,y,z) = \frac{\hbar}{m}\boldsymbol{\nabla}\theta_c(x,y,z) \ . \tag{3.273}$$

A single quantized vortex line at $(x=0, y=0)$ with topopogical charge $q \in \mathbb{Z}$ is obtained by setting

$$\theta_c(x,y,z) = q\ \arctan\left(\frac{y}{x}\right) = q\ \phi \ , \tag{3.274}$$

where $\phi = \arctan\left(\frac{y}{x}\right)$ is the polar angle and $r = \sqrt{x^2+y^2}$ is the polar radius of cylindrical coordinates. In this case, taking into account that $\boldsymbol{\nabla} = (\partial_x, \partial_y, \partial_z) = (\partial_r, \frac{1}{r}\partial_\phi, \partial_z)$ one finds

$$\mathbf{v}_c(r,\phi) = \frac{\hbar}{m}\frac{q}{r}\,\mathbf{u}_\phi = \frac{\hbar}{m}\frac{q}{\sqrt{x^2+y^2}}\Big(-\frac{y}{\sqrt{x^2+y^2}}, \frac{x}{\sqrt{x^2+y^2}}\Big)\,, \tag{3.275}$$

where $\mathbf{u}_\phi = \Big(-\frac{y}{\sqrt{x^2+y^2}}, \frac{x}{\sqrt{x^2+y^2}}, 0\Big)$ is a unit vector perpendicular to the position vector $(x, y, 0)$ at the angle ϕ. By solving the Gross-Pitaevskii equation one also finds

$$n_c(x,y) \simeq n_\infty \left(1 - \frac{1}{1 + \frac{x^2+y^2}{\xi^2}}\right) \tag{3.276}$$

where ξ is the healing length, namely the characteristic length as which the superfluid density $n_c(x,y)$ of the bosonic system is half of its bulk value n_∞. It is important to underline that quantized vortices have been observed experimentally in superfluid liquid helium, in ultracold atomic gases made of alkali-metal atoms, and also in type-II superconductors.

3.9.3 Bright and Dark Solitons

The real-time Gross-Pitaevskii equation (3.239) can be seen as the Euler-Lagrange equation associated to the following (real-time) "classical" action functional

$$\begin{aligned} S_c \;=\; & \int dt\, d^3\mathbf{r} \left\{ \psi_c^*(\mathbf{r},t) \left[i\hbar\frac{\partial}{\partial t} + \frac{\hbar^2}{2m}\nabla^2 - U(\mathbf{r}) \right] \psi_c(\mathbf{r},t) \right. \\ & \left. - \frac{1}{2}\, g\, |\psi_c(\mathbf{r},t)|^4 \right\} . \end{aligned} \tag{3.277}$$

Let us consider a very strong harmonic confinement of frequency $\omega_\perp$ along x and y, but no confinement along z, namely

$$U(\mathbf{r}) = \frac{1}{2} m\omega_\perp^2 (x^2 + y^2)\,. \tag{3.278}$$

We adopt the variational ansatz

$$\psi_c(\mathbf{r},t) = f(z,t) \frac{1}{\pi^{1/2}\ell_\perp^2} \exp\left(\frac{x^2+y^2}{2\ell_\perp^2}\right)\,, \tag{3.279}$$

where $f(z,t)$ is the axial wave function and $\ell_\perp = \sqrt{\hbar/(m\omega_\perp)}$ is the characteristic length of the transverse harmonic confinement. By inserting Eq. (3.279) into Eq. (3.277) and integrating along x and y, the resulting effective action functional depends on the field $f(z,t)$. One easily finds that the Euler-Lagrange equation of the axial wavefunction $f(z,t)$ reads

$$i\hbar \frac{\partial}{\partial t} f(z,t) = \left[-\frac{\hbar^2}{2m} \frac{\partial^2}{\partial z^2} + g_{1D} |f(z,t)|^2 \right] f(z,t) \, , \tag{3.280}$$

where

$$g_{1D} = \frac{g}{2\pi \ell_\perp^2} \tag{3.281}$$

is the effective one-dimensional interaction strength and the additive constant $\hbar\omega_\perp$ has been omitted because it does not affect the dynamics.

Eq. (3.280) is a one-dimensional cubic nonlinear Schrödinger equation. In 1972, Vladimir Zakharov and Aleksei Shabat found that this equation admits solitonic (i.e., shape invariant) analytical solutions. In particular, for $g_{1D} < 0$ (self-focusing nonlinearity) one finds the localized bright soliton solution

$$f(z,t) = \sqrt{\frac{m|g_{1D}|}{8\hbar^2}} \; sech \left[\frac{m|g_{1D}|}{4\hbar^2} (z - vt) \right] e^{iv(z-vt)} e^{i(mv^2/2 - \mu)t/\hbar} \, , \tag{3.282}$$

where $sech[x]$ is the hyperbolic secant, v is the arbitrary velocity of the propagation of the shape-invariant bright soliton and $\mu = -mg_{1D}^2/(16\hbar^2)$.

Instead, for $g_{1D} > 0$ (self-defocusing nonlinearity) one finds the localized but not normalized dark soliton solution

$$\begin{aligned} f(z,t) &= \sqrt{\frac{g_{1D}\phi_\infty^2 - mv^2}{g_{1D}}} \left(tanh \left[(z - vt) \sqrt{g_{1D}\phi_\infty^2 - v^2} \right] \right. \\ &+ \left. i \sqrt{\frac{mv^2}{g_{1D}\phi_\infty^2 - mv^2}} \right) e^{-i\mu t/\hbar} \, , \end{aligned} \tag{3.283}$$

where $tanh[x]$ is the hyperbolic tangent, $\phi_\infty > 0$ is the constant value of the field amplitude $|f(z,t)|$ at infinity ($z \to \pm\infty$), and v

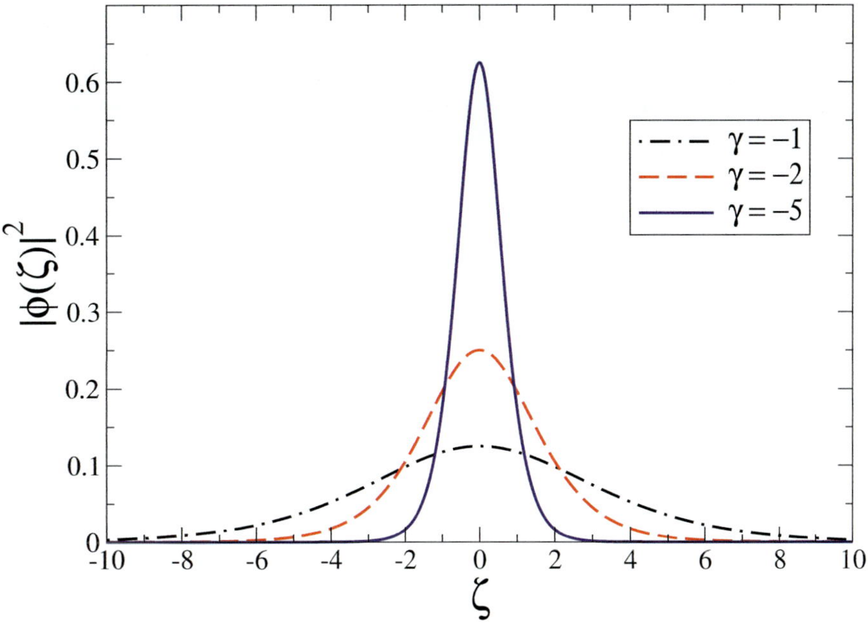

Figure 3.2
Probability density $|\phi(\zeta)|^2$ vs comoving coordinate $\zeta = z - vt$ of the bright soliton for three values of the nonlinear strength $\gamma = Ng_{1D} < 0$ with N the total number of attractive bosons. We set $\hbar = m = 1$.

is the velocity of propagation of the dark soliton. The parameter $\mu = g_{1D}\phi_\infty^2$ is obtained by the asymptotic behavior of the field $f(z,t)$. For $0 < |v| < c_s$, where $c_s = \sqrt{g_{1D}\phi_\infty^2}$ is the axial sound velocity (i.e., the 1D Bogoliubov velocity), the minimum value of the field $f(z,t)$ is greater than zero, and the wave is called gray soliton. If $v = 0$ (stationary dark soliton) the minimum of the field $f(z,t)$ is zero, and the wave is called a black soliton. The velocity c_s is the maximal velocity of the dark soliton, namely the velocity of a wave of infinitesimal amplitude, i.e., a sound wave, propagating in a medium of density $\rho_\infty = \phi_\infty^2$.

It is pedagogically very useful to prove that the 1D stationary Gross-Pitaevskii equation

$$\left[-\frac{\hbar^2}{2m}\frac{d^2}{dz^2} + g_{1D}|f(z)|^2\right] f(z) = \mu\, f(z) \ , \tag{3.284}$$

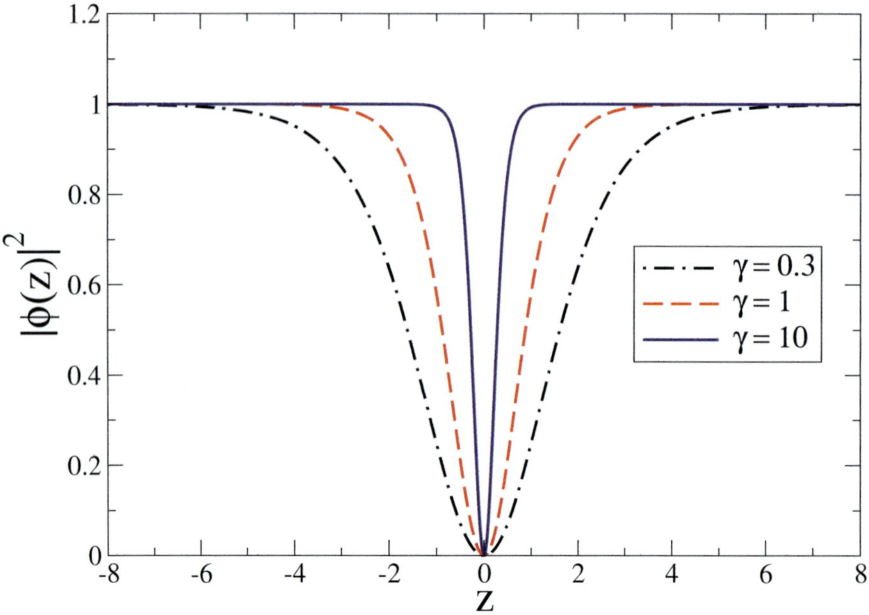

Figure 3.3
Probability density $|\phi(z)|^2$ vs coordinate z of the black soliton for three values of the nonlinear strength $\gamma = N g_{1D} > 0$, with N the total number of repulsive bosons. We set $\hbar = m = 1$.

with $g_{1D} < 0$ admits the stationary solution

$$f(z) = \sqrt{\frac{m|g_{1D}|}{8\hbar^2}} \, Sech \left[\frac{m|g_{1D}|}{4\hbar^2} z \right] \tag{3.285}$$

with

$$\mu = -\frac{m \; g_{1D}^2}{16 \; \hbar^2} \; . \tag{3.286}$$

To demonstrate this relevant result, let us assume that $f(z)$ is real. Then the 1D stationary Gross-Pitaevskii equation can be rewritten as

$$f''(z) = -\frac{\partial W(f)}{\partial f} \; , \tag{3.287}$$

where

$$W(f) = \frac{1}{2} \frac{m|g_{1D}|}{\hbar^2} f^4 + \frac{m\mu}{\hbar^2} f^2 \; . \tag{3.288}$$

Thus, $f(z)$ can be seen as the "coordinate" for a fictitious particle at "time" z. The constant of motion of the problem reads

$$K = \frac{1}{2} f'(z)^2 + W(f) \; , \tag{3.289}$$

from which one finds

$$\frac{df}{dz} = \sqrt{K - W(f)} \; . \tag{3.290}$$

Imposing that $f(z) \to 0$ as $|z| \to \infty$ one gets $K = 0$ and consequently

$$\frac{df}{\sqrt{-W(f)}} = dz \; , \tag{3.291}$$

or explicity

$$\frac{df}{\sqrt{-\frac{1}{2}\frac{m|g_{1D}|}{\hbar^2} f^4 + \frac{m|\mu|}{\hbar^2} f^2}} = dz \; , \tag{3.292}$$

with $\mu < 0$. Inserting the integrals, one obtains

$$\int_{f(0)}^{f(z)} \frac{df}{\sqrt{-\frac{1}{2}\frac{m|g_{1D}|}{\hbar^2} f^4 + \frac{m|\mu|}{\hbar^2} f^2}} = z \tag{3.293}$$

Setting $f'(0) = 0$, from the definition of K and using $K = 0$ one finds $W(f(0)) = 0$ and therefore

$$f(0) = \sqrt{\frac{2|\mu|}{|g_{1D}|}} \; . \tag{3.294}$$

After integration

$$\frac{1}{\sqrt{\frac{m|\mu|}{\hbar}}} ArcSech\left[\sqrt{\frac{|g_{1D}|}{2|\mu|}} f(z)\right] = z \tag{3.295}$$

from which

$$f(z) = \sqrt{\frac{2|\mu|}{|g_{1D}|}} Sech\left[\sqrt{\frac{m|\mu|}{\hbar^2}} z\right] \; . \tag{3.296}$$

Imposing the normalization condition

$$\int dz\ f(z)^2 = 1\ , \tag{3.297}$$

one obtains

$$\mu = -\frac{m\ g_{1D}^2}{16\ \hbar^2}\ . \tag{3.298}$$

3.9.4 Improved Dimensional Reduction

Bright and dark solitons have been obtained from the 1D Gross-Pitaevskii equation (1D GPE), which is derived from the 3D GPE assuming a transverse Gaussian with a constant width $\ell_\perp$. A more general assumption (Salasnich, Parola, and Reatto, 2002) is based on a space-time dependent transverse width

$$\psi_c(\mathbf{r},t) = f(z,t)\frac{1}{\pi^{1/2}\ell_\perp\sigma(z,t)}\exp\left(\frac{x^2+y^2}{2{\ell_\perp}^2\sigma(z,t)^2}\right)\ , \tag{3.299}$$

where $f(z,t)$ is the axial wave function and $\sigma(z,t)$ is the adimensional transverse width in units of $\ell_\perp$. We insert this ansatz in the 3D GP action functional (3.277) with the trapping potential

$$U(\mathbf{r}) = \frac{1}{2}m\omega_\perp^2(x^2+y^2) + \mathcal{U}(z) \tag{3.300}$$

with $\mathcal{U}(z)$ a generic potential along the z axis. Integrating along x and y, we find an effective 1D action functional which depends on the two fields $f(z,t)$ and $\sigma(z,t)$. Neglecting the spatial derivatives of $\sigma(z,t)$, the Euler-Lagrange equations associated to this effective 1D action are

$$i\hbar\frac{\partial}{\partial t}f = \left[-\frac{\hbar^2}{2m}\frac{\partial^2}{\partial z^2} + \mathcal{U}(z) + \frac{\gamma|f|^2}{\sigma^2} + \frac{\hbar\omega_\perp}{2}\left(\frac{1}{\sigma^2}+\sigma^2\right)\right]f\ , \tag{3.301}$$

$$\sigma = \left(1+\gamma|f|^2\right)^{1/4}\ , \tag{3.302}$$

where

$$\gamma = \frac{2a_S(N-1)}{\ell_\perp} \tag{3.303}$$

is the dimensional interaction strength. Inserting Eq. (3.302) into Eq. (3.301), we find

$$i\hbar\frac{\partial}{\partial t}f = \Big[-\frac{\hbar^2}{2m}\frac{\partial^2}{\partial z^2} + \mathcal{U}(z) + \frac{\gamma|f|^2}{\sqrt{1+\gamma|f|^2}} + \frac{\hbar\omega_\perp}{2}\left(\frac{1}{\sqrt{1+\gamma|f|^2}} + \sqrt{1+\gamma|f|^2}\right)\Big]f \quad (3.304)$$

that is the so-called 1D nonpolynomial Schrödinger equation (1D NPSE). This effective 1D wave equation describes the axial dynamics of a Bose condensate confined in an external potential with cylindrical symmetry. The trapping potential is harmonic in the transverse direction and generic in the axial one.

The 1D NPSE can be used to model cigar-shaped condensates, whose dynamics is essentially quasi-one-dimensional. In the weak-coupling regime $\gamma|f|^2 \ll 1$ one finds

$$\sigma \simeq 1 \quad (3.305)$$

and the 1D NPSE becomes the familiar 1D GPE

$$i\hbar\frac{\partial}{\partial t}f = \Big[-\frac{\hbar^2}{2m}\frac{\partial^2}{\partial z^2} + \mathcal{U}(z) + \gamma|f|^2 + \frac{\hbar\omega_\perp}{2}\Big]f\,. \quad (3.306)$$

However, in full generality, the 1D NPSE is much more reliable than the 1D GPE. With $\mathcal{U}(z) = 0$ and assuming $\gamma < 0$ the 1D NPSE admits analytical bright soliton solutions. Setting

$$f(z,t) = \phi(z - vt)e^{i(mv^2/2-\mu)t/\hbar}, \quad (3.307)$$

one finds the bright-soliton solution written in implicit form

$$\zeta = \frac{1}{\sqrt{2}}\frac{1}{\sqrt{1-\mu}}\,\mathrm{arctan}\left[\sqrt{\frac{\sqrt{1-|\gamma|\phi^2}-\mu}{1-\mu}}\right] \quad (3.308)$$

$$-\frac{1}{\sqrt{2}}\frac{1}{\sqrt{1+\mu}}\,\mathrm{arctanh}\left[\sqrt{\frac{\sqrt{1-|\gamma|\phi^2}-\mu}{1+\mu}}\right], \quad (3.309)$$

where $\zeta = z - vt$ and $|\gamma| = 2|a_s|(N-1)/\ell_\perp$. In the weak-coupling limit ($\gamma\phi^2 \ll 1$) one finds that the NPSE bright-soliton solution reduces the 1D GPE one. However, contrary to the 1D GPE bright soliton, the 1D NPSE bright soliton does not exist anymore, collapsing to a Dirac delta function at

$$\gamma_c = \left(\frac{2a_s(N-1)}{\ell_\perp}\right)_c = -\frac{4}{3} \, . \tag{3.310}$$

This analytical result is in very good agreement with the numerical solution of the 3D GPE but also with experimental results.

Quite remarkably, one can also deduce an effective 2D nonpolynomial Schrödinger equation (2D NPSE) that models disk-shaped Bose condensates confined in an external trap that is harmonic along the axial direction and generic in the transverse direction. It is important to stress that the 1D NPSE, and also the 2D NPSE, are routinely used by several experimental and theoretical groups to investigate BECs under strong confinement.

3.10 BKT Phase Transition

In three spatial dimensions (3D), the complex scalar field $\psi(\mathbf{r}, t)$ of superfluids is called the order parameter of the system, and it is often identified as the macroscopic wavefunction of Bose-Einstein condensation (BEC), where a macroscopic fraction of particles occupies the same single-particle quantum state. We have seen that for bosons in two spatial dimensions (2D), the BEC critical temperature is zero, while for bosons in one spatial dimension (1D) there is no BEC also at zero temperature. Despite the absence of BEC, in 1972 John Kosterlitz and David Thouless, with a previous contribution of Vadim Berezinskii, suggested that a 2D fluid can be superfluid below a critical temperature, the so-called Berezinskii-Kosterlitz-Thouless (BKT) critical temperature T_{BKT}. They analyzed the 2D XY model, which was originally used to describe the magnetization in a planar lattice of classical spins. Needless to say, the BKT phase transition has been observed experimentally with several 2D magnetic systems, thin films of superfluid ^{4}He, two-dimensional superconductors, and also with ultracold atomic gases in a quasi-2D configuration.

3.10.1 Continuous XY Model

The energy of the continuous 2D XY model is given by

$$E = \int d^2\mathbf{r}\, \frac{J_0}{2} \left(\nabla\theta\right)^2 \ , \tag{3.311}$$

where $\theta_c(\mathbf{r})$ is the angular field and J is the phase stiffness, also called phase rigidity. This is nothing else than the Schrödinger kinetic energy functional of the time-independent complex scalar field

$$\psi_c(\mathbf{r}) = \psi_0\, e^{i\theta_c(\mathbf{r})} \tag{3.312}$$

with a uniform modulus

$$\psi_0 = \sqrt{n_{s0}} \tag{3.313}$$

and

$$J_0 = \frac{\hbar^2 n_{s0}}{m} \ . \tag{3.314}$$

In fact, we have

$$\begin{aligned} E &= \int d^2\mathbf{r}\, \frac{\hbar^2}{2m} |\nabla\psi_c|^2 \\ &= \int d^2\mathbf{r}\, \frac{\hbar^2}{2m} \left\{ |\psi_c|^2 (\nabla\theta_c)^2 + \frac{1}{4|\psi_c|^2} (\nabla|\psi_c|^2)^2 \right\} \\ &= \int d^2\mathbf{r}\, \frac{\hbar^2 n_{s0}}{2m} (\nabla\theta_c)^2 \end{aligned} \tag{3.315}$$

because the quantum pressure term is zero under the assumption of a uniform configuration. Thus, the continuous 2D XY energy models the kinetic contribution of a superfluid described by a complex field $\psi(\mathbf{r})$.

A simple way to estimate the Berezinskii-Kosterlitz-Thouless (BKT) critical temperature T_{BKT} is to consider the Helmholtz free energy F of the 2D superfluid system characterized by one vortex line with topological charge $q = 1$ at temperature T. It is given by

$$F = E - T\,\mathcal{S}\,, \tag{3.316}$$

where the internal energy E reads

$$E = \pi J_0 \,\ln\left(\frac{R}{r_0}\right) \tag{3.317}$$

while the entropy $\mathcal{S}$ is given by

$$\mathcal{S} = k_B \ln\left(\frac{\pi R^2}{\pi r_0}\right) = 2k_B \ln\left(\frac{R}{r_0}\right)\,, \tag{3.318}$$

with $(\pi R^2)/(\pi r_0^2)$ the number of possible configurations of the vortex line of size r_0 in a circular domain of size R. It follows that

$$F = (\pi J_0 - 2k_B T)\,\ln\left(\frac{R}{r_0}\right)\,. \tag{3.319}$$

This function does not have singularities. The temperature T_c at which F changes sign is then

$$k_B T_c = \frac{\pi}{2} J_0 = \frac{\pi \hbar^2 n_{s0}}{2m}\,. \tag{3.320}$$

For $T < T_{BKT}$ the free energy F is positive, and in the limit $R \to +\infty$, it goes to $F \to +\infty$. Instead, $T > T_{BKT}$, the free energy F is negative, and in the limit $R \to +\infty$, it goes to $F \to -\infty$. Therefore, in the case of a very large system, T_{BKT} signals the topological phase transition from a superfluid phase ($0 \leq T < T_{BKT}$) without free vortices to a normal phase ($T > T_{BKT}$) characterized by the proliferation of free vortices.

3.10.2 Vortex-Antivortex Pairs

In the previous subsection we have discussed the simplest argument for the Kosterlitz-Thouless transition based on a single vortex. Actually, Eq. (3.320) gives the same critical temperature T_{BKT}

that one derives from a more sophisticated approach based on the formation of vortex-antivortex pairs as the temperature increases.

Let us consider a two-dimensional (2D) superfluid system characterized by two (effectively parallel along the z axis) vortex lines at the 2D positions $\mathbf{r}_1 = (x_1, y_1)$ and $\mathbf{r}_2 = (x_2, y_2)$ with opposite topological charges q and $-q$. This is the so-called vortex-antivortex configuration. The corresponding 2D superfluid velocity is given by

$$\mathbf{v}_c(\mathbf{r}) = \mathbf{v}_{s,1}(\mathbf{r}) + \mathbf{v}_{s,2}(\mathbf{r}) = \frac{\hbar}{m} \frac{q}{|\mathbf{r} - \mathbf{r}_1|} \mathbf{u}_1 - \frac{\hbar}{m} \frac{q}{|\mathbf{r} - \mathbf{r}_2|} \mathbf{u}_2 \ , \quad (3.321)$$

where

$$\mathbf{u}_j = \frac{1}{|\mathbf{r} - \mathbf{r}_j|} (y_j - y, x - x_j) \quad (3.322)$$

are 2D unit vectors perpendicular (in the plane XY) to the 2D position vectors $\mathbf{r} - \mathbf{r}_j = (x - x_j, y - y_j)$, with $j = 1, 2$. The kinetic energy of this system then reads

$$\begin{aligned} E &= \int \left\{ \frac{1}{2} m \; n_{s0} \; v_c(\mathbf{r})^2 \right\} \; d^2\mathbf{r} \\ &= \frac{1}{2} m \; n_{s0} \int \left\{ |\mathbf{v}_{s,1}(\mathbf{r}) + \mathbf{v}_{s,2}(\mathbf{r})|^2 \right\} \; d^2\mathbf{r} \\ &= \frac{\hbar^2 n_{s0} q^2}{2m} \int \left| \frac{\mathbf{u}_1}{|\mathbf{r} - \mathbf{r}_1|} - \frac{\mathbf{u}_2}{|\mathbf{r} - \mathbf{r}_2|} \right|^2 \; d^2\mathbf{r} \ . \end{aligned} \quad (3.323)$$

After some calculations one finds

$$E = E_1 + E_2 + E_{12} \ , \quad (3.324)$$

where

$$E_1 = E_2 = \frac{\pi \hbar^2 n_{s0}}{m} \; q^2 \; \ln \left(\frac{r_{max}}{r_0} \right) \quad (3.325)$$

$$E_{12} = 2\pi \frac{\hbar^2 n_{s0} q^2}{m} \ln \left(\frac{|\mathbf{r}_1 - \mathbf{r}_2|}{r_0} \right) . \quad (3.326)$$

The formation of a vortex-antivortex pair at a distance larger than r_0 has an energy cost. Thus, oppositely charged vortices attract, and they prefer to stay at the minimal distance, i.e., $|\mathbf{r}_1 - \mathbf{r}_2| = r_0$.

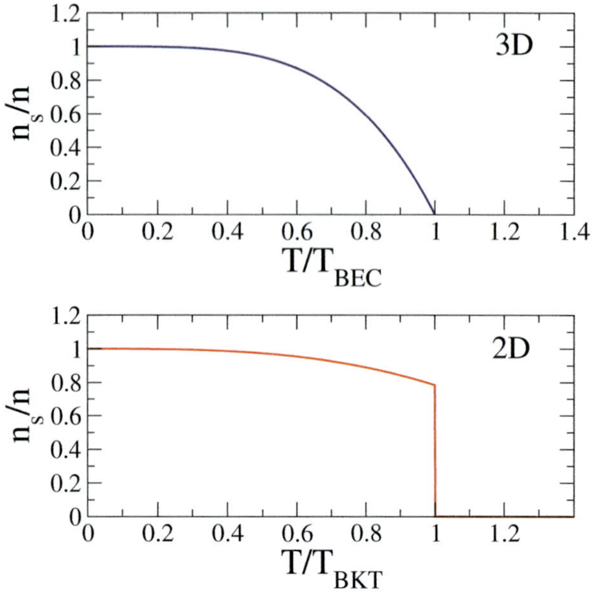

Figure 3.4
Superfluid fraction n_s/n as a function of temperature T in three spatial dimensions (3D) and in two-spatial dimensions (2D).

3.10.3 Renormalized Superfluid Density

The quite sophisticated analysis developed by Kosterlitz and Thouless shows that, as the temperature T increases, vortices start to appear in vortex-antivortex pairs (mainly with $q = \pm 1$). Moreover, the pairs are bound at low temperature until, at the critical temperature T_{BKT}, an unbinding transition occurs, above which a proliferation of free vortices and antivortices is predicted. Quite remarkably, the phase stiffness is renormalized by the presence of vortices: $J_0 \to J(T)$. The corresponding renormalized superfluid density

$$n_s(T) = J(T)\frac{m}{\hbar^2} \tag{3.327}$$

decreases by increasing the temperature T and jumps to zero above T_{BKT}. In Fig. 3.4 we show the key prediction of the Kosterlitz-Thouless transition: contrary to the 3D case, in 2D the superfluid fraction n_s/n jumps to zero above a critical temperature. For 3D superfluids the transition to the normal state is a BEC phase transition, while in 2D superfluids, the transition to the normal state is something different: a topological phase transition.

The main idea of Kosterlitz and Thouless is that the phase of the XY model

$$\theta_c(\mathbf{r}) = \theta_0(\mathbf{r}) + \theta_v(\mathbf{r}) \tag{3.328}$$

is decomposed as the sum of a vortex-free component $\theta_0(\mathbf{r})$ and component $\theta_{vor}(\mathbf{r})$ which instead contains all the quantized vortices of the system, i.e.

$$\theta_v(\mathbf{r}) = \sum_j q_j \arctan\left(\frac{y - y_i}{x - x_j}\right) \tag{3.329}$$

with $\mathbf{r}_j = (x_j, y_j)$ the position of the core of the quantized vortex with topological charge q_j. Notice that

$$(\nabla\theta_v(\mathbf{r}))^2 = \sum_{jk} q_j q_k \ln\left(\frac{|\mathbf{r}_j - \mathbf{r}_k|}{r_0}\right). \tag{3.330}$$

At temperature T the partition function of the system reads

$$\mathcal{Z} = \int \mathcal{D}[\theta_c] \sum_j e^{-\beta\int d^2\mathbf{r}\,\frac{J_0}{2}(\nabla\theta_c)^2} e^{\beta\mu_v q_j^2} \tag{3.331}$$

where $\beta = 1/(k_B T)$ and μ_{vor} is the chemical potential of the effective gas of quantized vortices. This grand canonical partition function of a gas of quantized vortices can be rewritten as

$$\mathcal{Z} = \int \mathcal{D}[\theta_0]\, e^{-\beta\int d^2\mathbf{r}\,\frac{J(T)}{2}(\nabla\theta_0)^2}. \tag{3.332}$$

It is possible to prove that the renormalized phase stiffness $J(T)$ can be obtained from the bare one J_0 through the Kosterlitz-Thouless renormalization group equations

$$\frac{\mathrm{d}}{\mathrm{d}\ell}K(\ell) = -4\pi^3 K(\ell)^2 y(\ell)^2 \tag{3.333}$$

$$\frac{\mathrm{d}}{\mathrm{d}\ell}y(\ell) = (2 - \pi K(\ell))\, y(\ell) \tag{3.334}$$

of the running variables $K(\ell)$ and $y(\ell)$, as a function of the adimensional scale ℓ. These running variables are subjected to the initial conditions $K(\ell=0)=J_0/(k_BT)$ and $y(\ell=0)=\exp(-\mu_v/(k_BT))$. The renormalized phase stiffness is then

$$J(T)=\frac{K(\ell=+\infty)}{k_BT}\ , \tag{3.335}$$

and the corresponding renormalized superfluid density $n_s(T)$ is given by Eq. (3.327). In the scientific literature the choice of the bare chemical potential μ_v is not unique. However, the numerical investigation of the quantized vortices by solving the 2D Gross-Pitaevskii equation suggests $\mu_v=\pi^2 J_0/4$ as the most rigorous choice for superfluids. Solving the renormalization group equations (3.333) and (3.334), one indeed finds that $n_s(T)$ jumps to zero at a critical temperature T_{BKT}. Moreover, one also derives the exact Nelson-Kosterlitz condition

$$k_BT_{BKT}=\frac{\hbar^2\pi}{8m}n_s(T_{BKT}^-)\ . \tag{3.336}$$

In some problems one is able to calculate the bare superfluid density $n_s^{(0)}(T)$ of the system, i.e., the finite-temperature superfluid density obtained without taking into account the presence of quantized vortices. For instance, in the case of 2D weakly-interacting bosons, adopting the Landau criterion we have

$$n_s^{(0)}(T)=n-\frac{1}{2k_BT}\frac{1}{2\pi^2}\int\frac{d^2q}{(2\pi)^2}\frac{\hbar^2q^2}{m}\frac{e^{\frac{E_k}{k_BT}}}{\left(e^{\frac{E_k}{k_BT}}-1\right)^2} \tag{3.337}$$

where E_k is the Bogoliubov spectrum. This formula can be used to estimate the BTK critical temperature T_{BKT}. In fact, we can write (Nelson-Kosterlitz criterion)

$$k_BT_{BKT}=\frac{\hbar^2\pi}{8m}n_s^{(0)}(T_{BKT}) \tag{3.338}$$

with the bare $n_s^{(0)}(T)$ instead of $n_s(T)$. In this way we get an approximated T_{BKT} without the effort of calculating the renormalized superfluid density $n_s(T)$.

3.11 Solved Problems

Problem 3.1

By using the Gaussian variational wavefunction

$$\psi(\mathbf{r}) = \frac{1}{\pi^{3/4}\ell_H^{3/2}\sigma^{3/2}} e^{-(x^2+y^2+z^2)/(2\ell_H^2\sigma^2)} ,$$

where $\ell_H = \sqrt{\hbar/(m\omega)}$ is the characteristic length of the harmonic confinement, and σ is the dimensionless variational parameter. Calculate the energy per particle of a Bose-Einstein condensate under harmonic confinement, given by

$$U(\mathbf{r}) = \frac{1}{2}m\omega^2(x^2+y^2+z^2) ,$$

taking into account that the Gross-Pitaevskii energy functional reads

$$\begin{aligned} E &= N\int d^3\mathbf{r}\left\{\psi^*(\mathbf{r})\left[-\frac{\hbar^2}{2m}\nabla^2 + \frac{1}{2}m\omega^2(x^2+y^2+z^2)\right]\psi(\mathbf{r}) \right. \\ &\quad \left. \times + \frac{1}{2}Ng|\psi(\mathbf{r})|^4\right\} . \end{aligned}$$

Solution

Inserting the Gaussian wavefunction in the Gross-Pitaevskii energy functional, after integration over spatial coordinates, we obtain the energy E as a function of σ:

$$E = N\hbar\omega\left(\frac{3}{4}\frac{1}{\sigma^2} + \frac{3}{4}\sigma^2 + \frac{\gamma}{2}\frac{1}{\sigma^3}\right) ,$$

where $\gamma = \sqrt{2/\pi}Na_s/\ell_H$ is the dimensionless strength of the inter-particle interaction with $g = 4\pi\hbar^2 a_s/m$ and a_s the s-wave scattering length. The best choice for σ is obtained by extremizing the energy function:

$$\frac{dE}{d\sigma} = 0 ,$$

from which we have

$$\sigma(\sigma^4 - 1) = \gamma \; .$$

It follows that σ grows with γ, if $\gamma > 0$. Instead, if $\gamma < 0$ there are two possible values of σ: one value corresponds to a minimum of the energy E (meta-stable solution), while the other value corresponds to a maximum of the energy E (unstable solution). It is then quite easy to show that these two solutions exist only if $\gamma > -4/5^{5/4}$.

Problem 3.2

Prove that for the nonlinear Schrödinger field $\psi(\mathbf{r},t)$ the quantity

$$N = \int_{\mathbb{R}^3} d^3\mathbf{r}\, |\psi(\mathbf{r},t)|^2$$

is a constant of motion, i.e., it is time independent. This quantity is usually called the number of particles.

Solution

The nonlinear Schrödinger field satisfies the time-evolution equation

$$i\hbar \frac{\partial}{\partial t}\psi(\mathbf{r},t) = \left[-\frac{\hbar^2}{2m}\nabla^2 + U(\mathbf{r}) + \int d^3\mathbf{r}'\, V(\mathbf{r},\mathbf{r}')|\psi(\mathbf{r}',t)|^2 \right] \psi(\mathbf{r},t) \; .$$

Its complex conjugate equation is

$$\begin{aligned} &- i\hbar \frac{\partial}{\partial t}\psi^*(\mathbf{r},t) \\ &\quad = \left[-\frac{\hbar^2}{2m}\nabla^2 + U(\mathbf{r}) + \int d^3\mathbf{r}'\, V(\mathbf{r},\mathbf{r}')|\psi(\mathbf{r}',t)|^2 \right] \psi^*(\mathbf{r},t) \; . \end{aligned}$$

Multiplying by ψ^* on the left of the first equation and by ψ on the left of the second equation, we find

$$\begin{aligned} i\hbar\psi^* \frac{\partial}{\partial t}\psi &= -\frac{\hbar^2}{2m}\psi^*\nabla^2\psi + U(\mathbf{r})|\psi|^2 + \int d^3\mathbf{r}' V(\mathbf{r},\mathbf{r}')|\psi(\mathbf{r}',t)|^2|\psi|^2 \\ -i\hbar\psi \frac{\partial}{\partial t}\psi^* &= -\frac{\hbar^2}{2m}\psi\nabla^2\psi^* + U(\mathbf{r})|\psi|^2 + \int d^3\mathbf{r}' V(\mathbf{r},\mathbf{r}')|\psi(\mathbf{r}',t)|^2|\psi|^2. \end{aligned}$$

Subtracting these two equations, we obtain

$$i\hbar\left(\psi^*\frac{\partial}{\partial t}\psi - \psi\frac{\partial}{\partial t}\psi^*\right) + \frac{\hbar^2}{2m}\left(\psi^*\nabla^2\psi - \psi\nabla^2\psi^*\right) = 0$$

or equivalently

$$\left(\psi^*\frac{\partial}{\partial t}\psi - \psi\frac{\partial}{\partial t}\psi^*\right) + \frac{\hbar}{2m\,i}\left(\psi^*\nabla^2\psi - \psi\nabla^2\psi^*\right) = 0 \; ,$$

that is

$$\frac{\partial}{\partial t}\left(\psi^*\psi\right) + \frac{\hbar}{2m\,i}\boldsymbol{\nabla}\cdot\left(\psi^*\boldsymbol{\nabla}\psi - \psi\boldsymbol{\nabla}\psi^*\right) = 0 \; .$$

This is nothing else than the continuity equation

$$\frac{\partial}{\partial t}n + \boldsymbol{\nabla}\cdot\mathbf{j} = 0.$$

having introduced the local number density

$$n(\mathbf{r},t) = |\psi(\mathbf{r},t)|^2 \; ,$$

such that

$$N = \int_{\mathbb{R}^3} d^3\mathbf{r}\, n(\mathbf{r},t) \; ,$$

and the current density

$$\mathbf{j}(\mathbf{r},t) = \frac{\hbar}{2m\,i}\left(\psi^*(\mathbf{r},t)\boldsymbol{\nabla}\psi(\mathbf{r},t) - \psi(\mathbf{r},t)\boldsymbol{\nabla}\psi^*(\mathbf{r},t)\right) \; .$$

Note that the current density can also be rewritten as

$$\mathbf{j}(\mathbf{r},t) = n(\mathbf{r},t)\,\mathbf{v}(\mathbf{r},t) \; ,$$

where

$$\mathbf{v}(\mathbf{r},t) = \frac{\hbar}{m}\boldsymbol{\nabla}\theta(\mathbf{r},t)$$

is the velocity field with $\theta(\mathbf{r},t)$ the angle of the Madelung representation

$$\psi(\mathbf{r},t) = \sqrt{n(\mathbf{r},t)}\, e^{i\theta(\mathbf{r},t)} \; .$$

Integrating over the full three-dimensional space, the continuity equation, we get

$$\int_{\mathbb{R}^3} d^3\mathbf{r}\, \frac{\partial}{\partial t} n(\mathbf{r}, t) + \int_{\mathbb{R}^3} d^3\mathbf{r}\, \boldsymbol{\nabla} \cdot \mathbf{j}(\mathbf{r}, t) = 0$$

which gives

$$\frac{\partial}{\partial t} \int_{\mathbb{R}^3} d^3\mathbf{r}\, n(\mathbf{r}, t) + \oint_{\partial\mathbb{R}^3} d^2\mathbf{r}\, \mathbf{j}(\mathbf{r}, t) \cdot \mathbf{u}(\mathbf{r}, t) = 0 \, .$$

Assuming that the current is zero at infinity

$$\frac{\partial}{\partial t} \int_{\mathbb{R}^3} d^3\mathbf{r}\, n(\mathbf{r}, t) = 0$$

namely

$$\frac{\partial}{\partial t} N = 0 \, .$$

Problem 3.3

Prove that for the nonlinear Klein-Gordon field $\varphi(\mathbf{r}, t)$ the quantity

$$Q = \int_{\mathbb{R}^3} d^3\mathbf{r} \left[\varphi^*(\mathbf{r}, t) \frac{\partial}{\partial t} \varphi(\mathbf{r}, t) - \varphi(\mathbf{r}, t) \frac{\partial}{\partial t} \varphi^*(\mathbf{r}, t) \right]$$

is a constant of motion, i.e., it is time independent. This quantity is usually called the total charge of particles and anti-particles.

Solution

The nonlinear Klein-Gordon field satisfies the time-evolution equation

$$\left(\frac{\partial^2}{\partial t^2} - \nabla^2 + \lambda |\varphi(\mathbf{r}, t)|^2 \right) \varphi(\mathbf{r}, t) = 0$$

Its complex conjugate equation is

$$\left(\frac{\partial^2}{\partial t^2} - \nabla^2 + \lambda |\varphi(\mathbf{r}, t)|^2 \right) \varphi^*(\mathbf{r}, t) = 0$$

Multiplying by φ^* on the left of the first equation and by φ on the left of the second equation, we find

$$\begin{aligned}\varphi^*\left(\frac{\partial^2}{\partial t^2}-\nabla^2\right)\varphi+\lambda|\varphi|^4 &= 0\\ \varphi\left(\frac{\partial^2}{\partial t^2}-\nabla^2\right)\varphi^*+\lambda|\varphi|^4 &= 0\,.\end{aligned}$$

Subtracting these two equations, we obtain

$$\left(\varphi^*\frac{\partial^2}{\partial t^2}\varphi-\varphi\frac{\partial^2}{\partial t^2}\varphi^*\right)+\left(\varphi^*\nabla^2\varphi-\varphi\nabla\varphi^*\right)=0$$

or equivalently

$$\frac{\partial}{\partial t}\left(\varphi^*\frac{\partial}{\partial t}\varphi-\varphi\frac{\partial}{\partial t}\varphi^*\right)+\boldsymbol{\nabla}\cdot\left(\varphi^*\boldsymbol{\nabla}\varphi-\varphi\boldsymbol{\nabla}\varphi^*\right)=0\,.$$

This is a sort of continuity equation

$$\frac{\partial}{\partial t}q+\boldsymbol{\nabla}\cdot\mathbf{j}=0\,,$$

where

$$q(\mathbf{r},t)=\varphi^*(\mathbf{r},t)\frac{\partial}{\partial t}\varphi(\mathbf{r},t)-\varphi(\mathbf{r},t)\frac{\partial}{\partial t}\varphi^*(\mathbf{r},t)$$

is the local charge density, such that

$$Q=\int_{\mathbb{R}^3}d^3\mathbf{r}\,\left[\varphi^*(\mathbf{r},t)\frac{\partial}{\partial t}\varphi(\mathbf{r},t)-\varphi(\mathbf{r},t)\frac{\partial}{\partial l}\varphi^*(\mathbf{r},t)\right]$$

is the total charge, and

$$\mathbf{j}(\mathbf{r},t)=\varphi^*(\mathbf{r},t)\boldsymbol{\nabla}\varphi(\mathbf{r},t)-\varphi(\mathbf{r},t)\boldsymbol{\nabla}\varphi^*(\mathbf{r},t)$$

is the local current density. Integrating over the full three-dimensional space the continuity equation, we get

$$\int_{\mathbb{R}^3}d^3\mathbf{r}\,\frac{\partial}{\partial t}q(\mathbf{r},t)+\int_{\mathbb{R}^3}d^3\mathbf{r}\,\boldsymbol{\nabla}\cdot\mathbf{j}(\mathbf{r},t)=0$$

which gives

$$\frac{\partial}{\partial t}\int_{\mathbb{R}^3} d^3\mathbf{r}\, q(\mathbf{r},t) + \oint_{\partial\mathbb{R}^3} d^2\mathbf{r}\, \mathbf{j}(\mathbf{r},t)\cdot\mathbf{u}(\mathbf{r},t) = 0 \, .$$

Assuming that the current is zero at infinity

$$\frac{\partial}{\partial t}\int_{\mathbb{R}^3} d^3\mathbf{r}\, q(\mathbf{r},t) = 0$$

namely

$$\frac{\partial}{\partial t} Q = 0 \, .$$

To conclude, we stress that, quite remarkably, in the case of the Klein-Gordon field, the quantity

$$\int_{\mathbb{R}^3} d^3\mathbf{r}\, |\varphi(\mathbf{r},t)|^2$$

is instead not a constant of motion.

Problem 3.4

Prove that

$$\prod_{n=-\infty}^{+\infty} (-i\,2\pi\,n + x)\, e^{i2\pi n0^+} = (1 - e^{-x}) \, ,$$

where the term $e^{i2\pi n0^+}$ is called causal regulator or convergence factor.

Solution

Let us set

$$\mathcal{P}(x) = \prod_{n=-\infty}^{+\infty} (-i\,2\pi\,n + x)\, e^{i2\pi n0^+} \, .$$

Then, taking into account that 0^+ is a positive infinitesimal, we find

$$\begin{aligned}
\ln(\mathcal{P}(x)) &= \sum_{n=-\infty}^{+\infty} \left[\ln(-i\,2\pi\,n + x) + i2\pi n0^+\right] \\
&= \sum_{n=-\infty}^{+\infty} \ln(-i\,2\pi\,n + x)\, e^{i2\pi n0^+}
\end{aligned}$$

and

$$\frac{d}{dx}\ln(\mathcal{P}(x)) = \sum_{n=-\infty}^{+\infty} \frac{e^{i2\pi n 0^+}}{-i\,2\pi\,n + x}\;.$$

We also set

$$\mathcal{Q}(x) = \sum_{n=-\infty}^{+\infty} \frac{e^{i2\pi n 0^+}}{-i\,2\pi\,n + x}$$

and develop a reasoning of complex analysis. Given the function

$$I_\delta(x) = \frac{1}{2\pi i}\oint_{\mathcal{C}} dz\,\frac{e^{z\delta}}{-z+x}\,\frac{1}{e^z - 1}$$

with $\mathcal{C}$ a very large (infinite) circle around the origin $z = 0$ in the Argand-Gauss plane of the complex variable z, we realize that the integrating function

$$f_\delta(z) = \frac{e^{z\delta}}{-z+x}\,\frac{1}{e^z - 1}$$

has a pole in the real axis at $z = x$ with residue $e^{x\delta}/(e^x - 1)$ but also poles in the imaginary axis at $z = i2\pi n$ with residues $1/(-i2\pi i + x)$. Moreover, $I_\delta(x)$ is such that

$$\lim_{\delta\to 0^+} I_\delta(x) = 0$$

because $f_\delta(z)$ is exponentially suppressed as $z \to \infty$ in the full Argand-Gauss plane. Here, it is crucial to have the convergence factor e^{z0^+}. In fact, without the term $e^{z\delta}$, i.e., setting $\delta = 0$ before the integration, one would get $f_0(z) \to 1/|z|$ for $\mathrm{Re}(z) \to -\infty$ and $I_0 = 1/2$ as a consequence. In this way one would obtain the results, we will derive in Problem 3.5 by adopting a different strategy. From the residue theorem applied to $I_\delta(x)$, we also have

$$\lim_{\delta\to 0^+} I_\delta(x) = \mathcal{Q}(x) - \frac{1}{e^x - 1}\;.$$

Thus, from the last two equations, we get

$$\mathcal{Q}(x) = \frac{1}{e^x - 1}\;.$$

Recalling that

$$\mathcal{Q}(x) = \frac{d}{dx} \ln(\mathcal{P}(x))$$

it follows that

$$\ln(\mathcal{P}(x)) = \int \mathcal{Q}(x)dx + K = \ln\bigl(1 - e^{-x}\bigr) + K$$

with K a constant of integration. Removing the constant K (which actually implies a regularization because K is infinite, but K does not depend on x), we finally obtain

$$\mathcal{P}(x) = \left(1 - e^{-x}\right) .$$

Problem 3.5
Prove that

$$\prod_{n=-\infty}^{+\infty} (-i\,2\pi\,n + x) = 2\,\sinh\Bigl(\frac{x}{2}\Bigr) .$$

Solution
We set

$$\mathcal{P}(x) = \prod_{n=-\infty}^{+\infty} (-i\,2\pi\,n + x) .$$

Then we find

$$\ln(\mathcal{P}(x)) = \sum_{n=-\infty}^{+\infty} \ln(-i\,2\pi\,n + x)$$

and also

$$\frac{d}{dx} \ln(\mathcal{P}(x)) = \sum_{n=-\infty}^{+\infty} \frac{1}{-i\,2\pi\,n + x} .$$

We set

$$\mathcal{Q}(x) = \sum_{n=-\infty}^{+\infty} \frac{1}{-i\,2\pi\,n + x}$$

noting that

$$\sum_{n=-\infty}^{+\infty} \frac{1}{-i\,2\pi\,n + x} = \sum_{n=-\infty}^{+\infty} \frac{1}{+i\,2\pi\,n + x} \; .$$

As a consequence

$$\begin{aligned} \mathcal{Q}(x) &= \sum_{n=-\infty}^{+\infty} \frac{1}{2} \left[\frac{1}{-i\,2\pi\,n + x} + \frac{1}{+i\,2\pi\,n + x} \right] \\ &= \sum_{n=-\infty}^{+\infty} \frac{1}{2} \left[\frac{2\,x}{(2\pi n)^2 + x^2} \right] = \sum_{n=-\infty}^{+\infty} \frac{x}{(2\pi n)^2 + x^2} \\ &= x \sum_{n=-\infty}^{+\infty} \frac{1}{(2\pi n)^2 + x^2} \; . \end{aligned}$$

The series converges and it is given by

$$\sum_{n=-\infty}^{+\infty} \frac{1}{(2\pi n)^2 + x^2} = \frac{1}{2x} \coth\left(\frac{x}{2}\right) \; .$$

Thus, we find

$$\mathcal{Q}(x) = \frac{1}{2} \coth\left(\frac{x}{2}\right) \; .$$

Recalling that

$$\mathcal{Q}(x) = \frac{d}{dx} \ln(\mathcal{P}(x))$$

it follows that

$$\ln(\mathcal{P}(x)) = \int \mathcal{Q}(x) dx + K = \ln\left(2 \sinh\left(\frac{x}{2}\right)\right) + K$$

with K a constant of integration. Regularizing the problem by removing the (infinite) constant K, we finally obtain

$$\mathcal{P}(x) = 2 \sinh\left(\frac{x}{2}\right) \; .$$

Further Reading

For a detailed discussion of the coherent state path integral in the case of the single harmonic oscillator:
– J.W. Negele and H. Orland, Quantum Many-particle Systems (Addison-Wesley, Boston, 1988).
– H. Kleinert, Path Integrals In Quantum Mechanics, Statistics, Polymer Physics, And Financial Markets (World Scientific, Singapore, 2009).
– G. Kordas, S. I. Mistakidis, and A. I. Karanikas, Coherent-state path integrals in the continuum, Phys. Rev. A **90**, 032104 (2014).
– A. Rancon, Hubbard-Stratonovich transformation and consistent ordering in the coherent state path integral: insights from stochastic calculus, J. Phys. A: Math. Theor. **53**, 105302 (2020).
For the functional integration of the non-relativistic bosonic field:
– N. Nagaosa, Quantum Field Theory in Condensed Matter Physics, chapt. 1, sections 1.1 and 1.2 (Springer, Berlin, 1999).
– H.T.C. Stoof, K.B. Gubbels, and D.B.M. Dickerscheid, Ultracold Quantum Fields, chapt. 6, sections 6.1, 6.2, 6.3, and 6.4 (Springer, Berlin, 2009).
– A. Altland and B. Simons, Condensed Matter Field Theory, chapt. 2, section 2.1 (Cambridge Univ. Press, Cambridge, 2006).
For a review of methods of regularization in the case of interacting bosons:
– L. Salasnich and F. Toigo, Zero-point energy of ultracold atoms, Phys. Rep. **640**, 1 (2016).
For bright and dark solitons in the nonlinear Schrödinger equation:
– A. Scott (ed.), Encyclopedia of Nonlinear Science (Taylor and Francis, New York, 2005).
The improved dimensional reduction for bosons described by the Gross-Pitaevskii equation is deeply analyzed here:
– L. Salasnich, A. Parola, and L. Reatto, Effective wave equations for the dynamics of cigar-shaped and disk-shaped Bose condensates, Phys. Rev. A **65**, 043614 (2002).

The Berezinskii-Kosterlitz-Thouless phase transition and its applications for superfluids, superconductors and magnetic systems is very well described in the book:

– J.V. Jose (ed.), 40 Years Of Berezinskii Kosterlitz Thouless Theory (World Scientific, Singapore, 2013).

4

Path Integral for Fermions

In this chapter we consider the quantum statistical mechanics of identical nonrelativistic fermions by introducing the formalism of functional integration for the fermionic field. We then derive the equation of state of non-interacting and weakly-interacting repulsive fermions. We also discuss the pairing of electrons within the mean-field Bardeen-Cooper-Schrieffer (BCS) theory of low-temperature superconductivity and the Ginzburg-Landau theory, which is quite successful near the critical temperature. Finally, we analyze the BCS-BEC crossover that was experimentally achieved with ultracold fermionic atoms twenty years ago and also Josephson junctions, which are the building blocks of superconducting quantum computers.

4.1 Introduction

For an ideal gas of identical fermions, we have seen in the first chapter that the distribution of the energies E is given by the Fermi-Dirac distribution

$$f(E) = \frac{1}{e^{(E-\mu)/(k_B T)} + 1} \,. \tag{4.1}$$

It is important to underline that if $(E-\mu)/k_B T \gg 1$, that is, at high temperatures, the Fermi-Dirac distribution reduces to the Maxwell-Boltzmann one

$$f(E) = z\, e^{-E/(k_B T)} \tag{4.2}$$

with $z = e^{\mu/(k_B T)}$ the fugacity. If instead $(E-\mu)/k_B T \ll 1$, i.e., at low temperatures, the behavior of the two distributions is very

DOI: 10.1201/9781003588337-4

different. At $T = 0$ (absolute zero) the distribution of Fermi-Dirac (4.1) becomes

$$f(E) = \Theta(E - \mu) , \tag{4.3}$$

where $\Theta(x)$ is the step function, which is 0 for $x < 0$ and 1 for $x > 0$. Therefore, at zero temperature, fermions have a distribution constant up to the energy μ, called the Fermi energy, while there cannot be fermions with a higher energy of the Fermi energy. At finite, but small temperature, only a fraction of fermions exceed the Fermi energy. Since the Fermi-Dirac distribution is equal to 1 for $E < \mu$, there can be at most one fermion for every allowed energy E of a single particle. This fact depends on the Pauli exclusion principle, according to which two fermions cannot have the same quantum numbers. Therefore, fermionic particles tend to stay separate from each other. Without the Pauli principle all atoms would be the same: it is the fermionic nature of the electrons in the atoms that gives rise to the complexity of the periodic table of elements by Dmitry Mendeleev.

Inside a material, electrons can, however, give rise to very special effects that, somehow, overcome the Pauli exclusion principle. In 1911 Heike Kamerlingh Onnes observed that the mercury, cooled at temperatures below 4.16 K, ceases suddenly to resist the passage of electric current: its resistance electricity becomes zero. Onnes, who received the Nobel Prize for his studies of physics in 1913, called this phenomenon superconductivity, and he discovered that it is not limited to mercury. Many materials below a critical temperature T_c allow the transportation of electric current without resistance. Among the elements that become superconductive are aluminum with $T_c = 1.19$ K and lead, which becomes superconducting at $T_c = 7.2$ K. Experiments reveal that the critical temperature T_c varies with changes of the mass M of the ions of the material according to the law $T_c = \gamma/\sqrt{M}$, where γ is a constant. This result, known as the isotopic effect, indicates the importance of the lattice crystalline for superconductivity. Copper, gold, and silver, that is, those materials that at room temperature are among the best electrical conductors, at low temperatures do not become superconductors. In fact, it does not yet exist a simple rule that allows you to establish a priori which materials can become superconductors at sufficiently low temperatures. We will show in the next sections of this chapter that one can think of superconductivity as superfluidity in the case of charged particles. During their motion, the electrons find the atoms that form the lattice on their path lens of the conductor, against which they can collide. Bouncing off atoms, electrons are deflected from their trajectory, and they lose some of the energy to the atoms themselves. This transfer of energy, which manifests itself through heating of the conductor, produces friction that opposes movement of electrons and causes the electrical resistance of normal metal. Below the critical temperature T_c a macroscopic number of electrons is able to move without resistance, i.e without losing energy due to collisions with the lattice crystalline, just like in superfluids.

Before discussing in detail the superconductivity, and their properties, consequences, and generalizations, in this chapter we will first develop the general path integral formulation for fermions.

4.2 Gaussian Integrals with Grassmann Variables

In this chapter we will analyze many properties of identical fermions described by a quantum field. For the fermionic quantum field, the functional integral can be explicitly calculated in the case of Gaussian integrals with Grassmann variables. A typical example is

$$\int d\bar{c}\, dc\, e^{-a\bar{c}\,c} = a \tag{4.4}$$

where c and $\bar{c}$ are the complex Grassmann integration variables, while a is a complex number. This formula is demonstrated taking into account the equivalence between the Grassmann-Berezin integration and the Grassmann differentiation. In particular, we have

$$\int d\bar{c}\, dc\, e^{-a\bar{c}c} = \int d\bar{c}dc\ (1 - a\bar{c}c) = \frac{\partial^2}{\partial\bar{c}\partial c} 1 + a \frac{\partial^2}{\partial\bar{c}\partial c} c\,\bar{c} = a\ . \tag{4.5}$$

Another useful Gaussian-Grassmann-Berezin integral is

$$\int d\bar{c}\, dc\, e^{-a\bar{c}c+b\bar{c}+\bar{b}c} = a\, e^{\bar{b}\, a^{-1}\, b} \tag{4.6}$$

where a is a complex number and b is a complex Grassmann number. This result is proved by performing the change of variables

$$\begin{aligned} c &= c' + a^{-1}b \quad (4.7)\\ \bar{c} &= \bar{c}' + a^{-1}\bar{b} \quad (4.8) \end{aligned}$$

which gives $dc = dc'$, $d\bar{c} = d\bar{c}'$ and also

$$\int d\bar{c}\, dc\, e^{-a\bar{c}c+b\bar{c}+\bar{b}c} = \int d\bar{c}'\, dc'\, e^{-\bar{c}\, a^{-1}\, c}\, e^{\bar{b}\, a^{-1}\, b} = a\, e^{\bar{b}\, a^{-1}\, b}\,. \tag{4.9}$$

For Gaussian integrals with N complex Grassmann variables c_j $(j = 1, 2, ..., N)$, given the $N \times N$ hermitian matrix $\hat{A}$ and the vector $\mathbf{b}$ with N complex Grassmann components, one can obtain

$$\int \prod_{j=1}^{N} d\bar{c}_j\, dc_i\, e^{-\bar{\mathbf{c}}\cdot\hat{A}\mathbf{c}+\mathbf{b}\cdot\bar{\mathbf{c}}+\bar{\mathbf{b}}\cdot\mathbf{c}} = \det[\hat{A}]\, e^{\bar{\mathbf{c}}\cdot\hat{A}^{-1}\mathbf{c}}\,, \tag{4.10}$$

where $\det[\hat{A}]$ is the determinant of the matrix $\hat{A}$ and $\hat{A}^{-1}$ is the inverse of the matrix $\hat{A}$. It is important to stress that $\det[\hat{A}]$ appears in the numerator, while it would appear in the denominator treating bosonic variables.

4.3 Coherent State Path Integral: Fermionic Oscillator

Let us consider a single fermionic mode with annihilation operator $\hat{c}$ and creation operator $\hat{c}^+$, satisfying the anti-commutation relations

$$\begin{aligned} \{\hat{c}, \hat{c}^+\} &= 1 \quad (4.11)\\ \{\hat{c}, \hat{c}^+\} &= \{\hat{c}, \hat{c}^+\} = 0 \quad (4.12) \end{aligned}$$

with $\{A, B\} = \hat{A}\hat{B} + \hat{B}\hat{A}$ the anti-commutator. As usual $\hat{N} = \hat{c}^+\hat{c}$ is the single-mode number operator. See the first chapter for further details. The Hamiltonian for this fermionic mode can be written as:

$$\hat{H} = \epsilon\, \hat{c}^+\hat{c} = \epsilon\, \hat{N} \,, \tag{4.13}$$

where ϵ is the energy of the fermionic state. This Hamiltonian can be interpreted as the Hamiltonian of a sort of fermionic harmonic oscillator with frequency $\omega = \epsilon/\hbar$. For the sake of simplicity, we do not insert any zero-point energy in this fermionic case. The canonical partition function of the problem is easily obtained by using the two Fock states $|0\rangle$ and $|1\rangle$, namely

$$\mathcal{Z}_1 = \mathrm{Tr}[e^{-\beta\hat{H}}] = \sum_{n=0,1} \langle n|e^{-\beta\epsilon\,\hat{N}}|n\rangle = 1 + e^{-\beta\epsilon} \,. \tag{4.14}$$

The procedure to compute the partition function $\mathcal{Z}_1$ adopting an approach based on the fermionic coherent state path integral is much more involved. First we observe that, given the fermionic coherent state $|c\rangle$, that is the eigenstate of the fermionic annihilation operator $\hat{c}$ with Grassmann eigenvalue c, i.e.

$$\hat{c}|c\rangle = c|c\rangle \qquad \text{and} \qquad \langle c|\hat{c} = \bar{c}\langle c| \,, \tag{4.15}$$

the trace of an operator $\hat{A}$, written in terms of pairs of fermionic ladder operators $\hat{c}$ and $\hat{c}^+$, is given by

$$\mathrm{Tr}[\hat{A}] = \int d\bar{c}\; dc\, e^{-\bar{c}c}\, \langle -c|\hat{A}|c\rangle \,. \tag{4.16}$$

In fact, remembering the properties of Grassmann numbers and fermionic coherent states discussed in Chapter 1, we have

$$
\begin{aligned}
&\int d\bar{c}\, dc\, e^{-\bar{c}c}\, \langle -c|\hat{A}|c\rangle \\
&\quad = \int d\bar{c}\, dc\, (1-\bar{c}c)\, (\langle 0| + \bar{c}\langle 1|)\hat{A}(|0\rangle - c|1\rangle) \\
&\quad = \int d\bar{c}\, dc\, (1-\bar{c}c)\, \Big(\langle 0|\hat{A}|0\rangle - c\langle 0|\hat{A}|1\rangle + \bar{c}\langle 1|\hat{A}|0\rangle - \bar{c}c\langle 1|\hat{A}|1\rangle\Big) \\
&\quad = \int d\bar{c}\, dc\, \Big(\langle 0|\hat{A}|0\rangle - c\langle 0|\hat{A}|1\rangle + \bar{c}\langle 1|\hat{A}|0\rangle - \bar{c}c\langle 1|\hat{A}|1\rangle \\
&\qquad -\bar{c}c\langle 0|\hat{A}|0\rangle\Big) \\
&\quad = \int d\bar{c}\, dc\, \Big(c\bar{c}\langle 1|\hat{A}|1\rangle + c\bar{c}\langle 0|\hat{A}|0\rangle\Big) \\
&\quad = \langle 0|\hat{A}|0\rangle + \langle 1|\hat{A}|1\rangle\ .
\end{aligned}
\tag{4.17}
$$

Here, we have assumed that $\hat{A}$ commutes with Grassmann numbers, i.e., it contains the pair $\hat{c}^+\hat{c}$ as happens for the Hamiltonian operator $\hat{H}$. Please, observe that in the fermionic coherent state representation of the trace $\mathrm{Tr}[\hat{A}]$ it appears $\langle -c|$ instead of $\langle c|$. As we will see, this has a consequence on the fermionic path integral.

To develop the fermionic path integral, we must repeat the machinery discussed in previous chapters. We start from

$$
\begin{aligned}
\mathcal{Z}_1 &= \mathrm{Tr}[e^{-\beta\hat{H}}] = \int d\bar{c}_0\, dc_0\, e^{-\bar{c}_0 c_0}\, \langle -c_0|e^{-\beta\hat{H}}|c_0\rangle \\
&= \int d\bar{c}_N\, dc_N\, e^{-\bar{c}_N c_N}\, \langle c_N|e^{-\beta\hat{H}}|c_0\rangle
\end{aligned}
\tag{4.18}
$$

imposing

$$
c_0 = -c_N \qquad \text{and} \qquad \bar{c}_0 = -\bar{c}_N\ . \tag{4.19}
$$

Then we set $\Delta\tau = \hbar\beta/N$ and consider the Trotter decomposition

$$
e^{-\beta\hat{H}} = e^{-\frac{\Delta\tau}{\hbar}N\hat{H}} \simeq \left(e^{-\frac{\Delta\tau}{\hbar}\hat{H}}\right)^N = e^{-\frac{\Delta\tau}{\hbar}\hat{H}} \ldots e^{-\frac{\Delta\tau}{\hbar}\hat{H}} \tag{4.20}
$$

which becomes exact for $\Delta\tau \to 0$ and $N \to \infty$. Subsequently, we insert for $N-1$ times into Eq. (4.18) the spectral representation of the identity

$$
\hat{1} = \int d\bar{c}_j\, dc_j\, e^{-\bar{c}_j c_j}\, |c_j\rangle\langle c_j| \tag{4.21}
$$

with fermionic coherent states $|c_j\rangle$, $i = 1, 2, ..., N-1$. It follows that

$$\mathcal{Z}_1 \simeq \int \left(\prod_{j=1}^{N} d\bar{c}_j dc_j \right) e^{-\sum_{j=1}^{N} \bar{c}_j c_j} \langle c_N | e^{-\frac{\Delta\tau}{\hbar}\hat{H}} | c_{N-1} \rangle \; ... \; \langle c_1 | e^{-\frac{\Delta\tau}{\hbar}\hat{H}} | c_0 \rangle .$$

Keeping in mind the formula

$$\langle c_j | c_{j-1} \rangle = e^{\bar{c}_j \, c_{j-1}} = 1 + \bar{c}_j \, c_{j-1} \tag{4.22}$$

and also

$$\begin{aligned} &\langle c_j | e^{-\frac{\Delta\tau}{\hbar}\hat{H}} | c_{j-1} \rangle \\ &\quad \simeq \langle c_j | 1 - \frac{\Delta\tau}{\hbar} \hat{H} | c_{j-1} \rangle = \langle c_j | c_{j-1} \rangle - \langle c_j | \frac{\Delta\tau}{\hbar} \, \epsilon \, \hat{c}^{+} \, \hat{c} | c_{j-1} \rangle \\ &\quad = \langle c_j | c_{j-1} \rangle \left(1 - \frac{\Delta\tau}{\hbar} \, \epsilon \, \bar{c}_j \, c_{j-1} \right) = e^{\bar{c}_j \, c_{j-1} - \frac{\Delta\tau}{\hbar} \, \epsilon \, \bar{c}_j \, c_{j-1}} \, , \end{aligned} \tag{4.23}$$

after some lengthy manipulations we get

$$\mathcal{Z}_1 \simeq \int \left(\prod_{j=1}^{N} d\bar{c}_j \, dc_j \right) \; e^{-\frac{1}{\hbar} \sum_{j=1}^{N} [\hbar \bar{c}_j \frac{(c_j - c_{j-1})}{\Delta\tau} + \epsilon \, \bar{c}_j \, c_{j-1}] \Delta\tau} \tag{4.24}$$

In the continuum limit $\Delta \to 0$, with $\tau_j = j\Delta\tau$, and $c_j = c(\tau_j)$, we obtain a Grassmann-valued function $c(\tau)$ with $\tau \in [0, \hbar\beta]$, and the canonical partition function can be expressed as

$$\mathcal{Z}_1 = \int \mathcal{D}[\bar{c}, c] \, e^{-\frac{1}{\hbar} S_E[\bar{c}, c]} \, , \tag{4.25}$$

where $S_E[\bar{c}, c]$ is the Euclidean action

$$S_E[\bar{c}, c] = \int_0^{\hbar\beta} d\tau \, \bar{c}(\tau) \left(\hbar \frac{\partial}{\partial\tau} + \epsilon \right) c(\tau) \tag{4.26}$$

and

$$\int \mathcal{D}[\bar{c}, c] = \int \prod_{\tau} d\bar{c}(\tau) dc(\tau) \, . \tag{4.27}$$

Remember that $c(\tau)$ is the Grassmann dynamical variable and $\bar{c}(\tau)$ is its conjugate Grassmann dynamical variable. Here, the integrals are Grassmann integrals. To properly represent fermions, we have imposed anti-periodic boundary conditions, namely

$$c(0) = -c(\hbar\beta), \qquad \bar{c}(0) = -\bar{c}(\hbar\beta) \,. \tag{4.28}$$

If we use periodic boundary conditions, the dynamical variables would behave like bosons, leading to an incorrect result for the partition function. Under anti-periodic boundary conditions, the fermionic mode acquires a minus sign when looping around the imaginary time interval. This reflects the anti-commuting nature of fermionic operators, in particular the presence of $\langle -c|$ in the trace of Eq. (4.16), and ensures that the correct fermionic statistics are encoded.

The Grassmann dynamical variable $c(\tau)$ can be expanded in terms of Fourier modes as

$$c(\tau) = \sum_{n=-\infty}^{+\infty} c_n \, e^{-i\omega_n \tau} \,, \tag{4.29}$$

where the frequencies ω_n must be odd multiples of π/β due to the anti-periodic boundary conditions, namely

$$\omega_n = \frac{(2n+1)\pi}{\hbar\beta} \,. \tag{4.30}$$

These frequencies are called fermionic Matsubara frequencies. Notice that one can also write $\omega_n = n\pi/(\hbar\beta)$ with odd n. This equivalent definition is sometimes useful because with it one has $\omega_{-n} = -\omega_n$, as in the bosonic case. Substituting this expansion into the Euclidean action, we find

$$S_E[\bar{c}, \psi] = \hbar \sum_n \bar{c}_n [\beta \, (-i\hbar\omega_n + \epsilon)] c_n \tag{4.31}$$

and the partition function becomes

$$\mathcal{Z}_1 = \prod_{n=-\infty}^{+\infty} \int d\bar{c}_n dc_n \, e^{-\bar{c}_n[\beta(-i\hbar\omega_n + \epsilon)]c_n} = \prod_{n=-\infty}^{+\infty} \beta(-i\hbar\omega_n + \epsilon) \,. \tag{4.32}$$

The last step requires to write

$$\mathcal{Z}_1 = \prod_{n=-\infty}^{+\infty} \beta(-i\hbar\omega_n + \epsilon) e^{i\omega_n 0^+} \tag{4.33}$$

where the convergence factor $e^{i\omega_n 0^+}$ is introduced, as in the bosonic case, to have the correct time ordering, i.e., the same as the discrete path integral. Then we must use

$$\prod_{n=-\infty}^{+\infty} \beta(-i\hbar\omega_n + \epsilon) e^{i\omega_n 0^+} = 1 + e^{\beta\epsilon} , \tag{4.34}$$

whose detailed derivation is given in Problem 4.3. In Problem 4.4 we calculate instead the product over fermionic Matsubara frequencies without the convergence factor: the final result is slightly different and contains the fermionic zero-point energy. Finally, using Eqs. (4.33) and (4.34) we get

$$\mathcal{Z}_1 = 1 + e^{-\beta\epsilon} , \tag{4.35}$$

that is the wanted result.

4.3.1 Fermionic Zero-Point Energy

It is useful to say a few words about the zero-point energy of the fermionic oscillator. We have used the following Hamiltonian for the single-mode oscillator written in terms of Grassmann variables:

$$H = \epsilon\, \bar{c}\, c \, . \tag{4.36}$$

Taking into account that $c\,\bar{c} = -\bar{c}\,c$, the fully symmetric version of the same Hamiltonian reads

$$H = \frac{\epsilon}{2} \left(\bar{c}\,c - c\,\bar{c} \right) \, . \tag{4.37}$$

Under the assumption that this is the appropriate classical-like analog of the quantum Hamiltonian, promoting the Grassmann variables to fermionic operators, i.e.

$$c \to \hat{c} \qquad \text{and} \qquad \bar{c} \to \hat{c}^+ , \tag{4.38}$$

we get the Hamiltonian operator

$$\hat{H} = \frac{\epsilon}{2}\left(\hat{c}^{+}\,\hat{c} - \hat{c}\,\hat{c}^{+}\right) . \tag{4.39}$$

The fermionic ladder operators are such that $\hat{c}\hat{c}^{+} = 1 - \hat{c}^{+}\hat{c}$, and consequently, we can write

$$\hat{H} = \epsilon\left(\hat{c}^{+}\hat{c} - \frac{1}{2}\right) = \epsilon\,\hat{N} - \frac{\epsilon}{2} \tag{4.40}$$

which displays a negative zero-point energy $-\epsilon/2$.

Another way to obtain the same result is to consider both the commutator $[\hat{A}, B] = \hat{A}\hat{B} - \hat{B}\hat{A}$ and the anti-commutator $\{\hat{A}, \hat{B}\} = \hat{A}\hat{B} + \hat{A}\hat{B}$ of two generic operators $\hat{A}$ and $\hat{B}$. Quite remarkably, the bosonic oscillator can be written in terms of the anti-commutator

$$\hat{H}_{\text{bosonic}} = \epsilon\,\{\hat{c}^{+}, \hat{c}\} = \epsilon\,\hat{c}^{+}\,\hat{c} + \frac{\epsilon}{2} \tag{4.41}$$

with ladder operators satisfying bosonic commutation rules, i.e., $[\hat{c}, \hat{c}^{+}] = 1$. It is then somehow natural to write the fermionic oscillator in terms of the commutator

$$\hat{H}_{\text{fermionic}} = \epsilon\,[\hat{c}^{+}, \hat{c}] = \epsilon\,\hat{c}^{+}\,\hat{c} - \frac{\epsilon}{2} \tag{4.42}$$

with ladder operators satisfying fermionic anti-commutation rules, i.e., $\{\hat{c}, \hat{c}^{+}\} = 1$. This kind of duality between bosons and fermions is the key ingredient in supersymmetric theories.

4.4 Partition Function of Interacting Fermions

For a system of non-relativistic interacting two-spin-component fermions, the partition function is given by

$$\mathcal{Z} = \text{Tr}[e^{-\beta(\hat{H} - \mu_{\uparrow}\hat{N}_{\uparrow} - \mu_{\downarrow}\hat{N}_{\downarrow})}] , \tag{4.43}$$

where the quantum-field-theory Hamiltonian is

$$
\begin{aligned}
\hat{H} &= \sum_{\sigma=\uparrow,\downarrow} \int d^3\mathbf{r}\ \hat{\psi}_\sigma^+(\mathbf{r})\Big(-\frac{\hbar^2}{2m}\nabla^2 + U_\sigma(\mathbf{r})\Big)\hat{\psi}_\sigma(\mathbf{r}) \\
&+ \frac{1}{2}\sum_{\sigma,\sigma'=\uparrow,\downarrow} \int d^3\mathbf{r}\ d^3\mathbf{r}'\ \hat{\psi}_\sigma^+(\mathbf{r})\ \hat{\psi}_{\sigma'}^+(\mathbf{r}')\ V_{\sigma,\sigma'}(\mathbf{r},\mathbf{r}')\ \hat{\psi}_{\sigma'}(\mathbf{r}')\ \hat{\psi}_\sigma(\mathbf{r})\ ,
\end{aligned}
\tag{4.44}
$$

with $U_\sigma(\mathbf{r})$ the external potential acting on fermions with spin σ, $V_{\sigma,\sigma'}(\mathbf{r},\mathbf{r}')$ the inter-particle interaction, while the number operator for fermions with spin σ reads

$$
\hat{N}_\sigma = \int d^3\mathbf{r}\ \hat{\psi}_\sigma^+(\mathbf{r})\hat{\psi}_\sigma(\mathbf{r})\ . \tag{4.45}
$$

As usual, $\beta = 1/(k_B T)$ with k_B the Boltzmann constant and T the temperature, while μ_σ is the chemical potential for the fermions with spin σ. The Hamiltonian (4.44) generalizes the one, valid for spinless fermions, described in the first chapter. Assuming that i) the external confinement does not depend on spin σ, ii) there is a contact interaction of strength g between fermions, and iii) there is chemical equilibrium, i.e., $\mu_\uparrow = \mu_\downarrow = \mu$, we can write

$$
\mathcal{Z} = \mathrm{Tr}[e^{-\beta(\hat{H}-\mu\hat{N})}] \tag{4.46}
$$

with

$$
\hat{N} = \hat{N}_\uparrow + \hat{N}_\downarrow \tag{4.47}
$$

and

$$
\begin{aligned}
\hat{H} &= \sum_{\sigma=\uparrow,\downarrow} \int d^3\mathbf{r}\ \hat{\psi}_\sigma^+(\mathbf{r})\Big(-\frac{\hbar^2}{2m}\nabla^2 + U(\mathbf{r})\Big)\hat{\psi}_\sigma(\mathbf{r}) \\
&+ g\int d^3\mathbf{r}\ \hat{\psi}_\uparrow^+(\mathbf{r})\ \hat{\psi}_\downarrow^+(\mathbf{r})\ \hat{\psi}_\downarrow(\mathbf{r})\ \hat{\psi}_\uparrow(\mathbf{r})\ .
\end{aligned}
\tag{4.48}
$$

Notice that in the presence of a contact inter-particle potential, there is an interaction only between fermions with opposite spins. Formally, one obtains Eq. (4.48) setting $V_{\sigma,\sigma'}(\mathbf{r},\mathbf{r}') = g\delta^{(3)}(\mathbf{r} -$

$\mathbf{r}')\,(1-\delta_{\sigma,\sigma'})$ in Eq. (4.44) with g the strength of the contact interparticle interaction potential.

In the previous chapter we have seen that the partition function of identical bosons can be expressed as a functional integral, taking into account that the trace of an operator can be performed by using any complete set of basis states, and, in particular, the basis of generalized coherent states of bosons. The same reasoning can be applied to the partition function of identical fermions by using the basis of generalized coherent states of fermions. As shown in the first chapter, in the case of fermionic coherent states, one must use anti-commuting complex Grassmann variables and anti-commuting complex Grassmann fields. Thus, we can write

$$\hat{\psi}_\sigma(\mathbf{r})\,|\psi\rangle = \psi_\sigma(\mathbf{r})\,|\psi\rangle\ , \tag{4.49}$$

where $\psi_\sigma(\mathbf{r})$ is the complex Grassmann field with spin σ and $|\psi\rangle$ is the generalized fermionic coherent state. The complex conjugate of the Grassmann field $\psi_\sigma(\mathbf{r})$ is denoted as $\bar{\psi}_\sigma(\mathbf{r})$ and it satisfies the equation

$$\langle\psi|\hat{\psi}^+_\sigma(\mathbf{r}) = \langle\psi|\bar{\psi}_\sigma(\mathbf{r})\ . \tag{4.50}$$

By expanding the fermionic field operator $\hat{\psi}_\sigma(\mathbf{r})$ as

$$\hat{\psi}_\sigma(\mathbf{r}) = \sum_\alpha \hat{c}_{\alpha,\sigma}\,\phi_\alpha(\mathbf{r})\ , \tag{4.51}$$

where $\phi_\alpha(\mathbf{r})$ are orthonormal eigenfunctions of single-particle states $|\alpha\rangle$ and $\hat{c}_{\alpha,\sigma}$ is the fermionic operator that annihilates a fermion with spin σ in the state $|\alpha\rangle$, we have

$$\psi_\sigma(\mathbf{r}) = \sum_\alpha c_{\alpha,\sigma}\,\phi_\alpha(\mathbf{r}) \tag{4.52}$$

with

$$\hat{c}_{\alpha,\sigma}|\psi\rangle = c_{\alpha,\sigma}|\psi\rangle \qquad \text{for any } \alpha \text{ and } \sigma =\uparrow,\downarrow \tag{4.53}$$

and $c_{\alpha,\sigma}$ the complex Grassmann variables for the single-particle state $|\alpha\rangle$ and spin σ.

Making a long story short, in analogy with the bosonic case, the partition function of identical fermions described by the many-body Hamiltonian (4.48) can be written as

$$\mathcal{Z} = \int \mathcal{D}[\bar{\psi}_\sigma, \psi_\sigma] \; \exp\left\{ -\frac{1}{\hbar} \, S_E[\bar{\psi}_\sigma, \psi_\sigma] \right\} , \qquad (4.54)$$

where

$$S_E[\bar{\psi}_\sigma, \psi_\sigma] = \int_0^{\hbar\beta} d\tau \int d^3\mathbf{r} \; \mathscr{L}_E(\psi_s, \bar{\psi}_s) \qquad (4.55)$$

is the Euclidean action functional and

$$\mathscr{L}_E = \sum_{\sigma=\uparrow,\downarrow} \bar{\psi}_\sigma \left[\hbar\partial_\tau - \frac{\hbar^2}{2m}\nabla^2 + U(\mathbf{r}) - \mu \right] \psi_\sigma + g \, \bar{\psi}_\uparrow \, \bar{\psi}_\downarrow \, \psi_\downarrow \, \psi_\uparrow \qquad (4.56)$$

is the Euclidean Lagrangian density, where $\psi_\sigma(\mathbf{r}, \tau)$ and $\bar{\psi}_\sigma(\mathbf{r}, \tau)$ are the Grassmann fields which also depend on the imaginary time τ. Notice that one picks up a minus sign each time one permutes two Gaussian fields. In the fermionic functional integral (4.54), the integration measure reads

$$\int \mathcal{D}[\bar{\psi}_\sigma, \psi_\sigma] = \int \prod_{\sigma=\uparrow,\downarrow} \prod_{(\mathbf{r},\tau)} d\bar{\psi}_\sigma(\mathbf{r}, \tau) \, d\psi_\sigma(\mathbf{r}, \tau) \; , \qquad (4.57)$$

and the boundary conditions for the imaginary time τ are anti-periodic, i.e., $\psi(\mathbf{r}, 0) = -\psi(\mathbf{r}, \hbar\beta)$. These boundary conditions ensure that the path integral respects the fermionic anti-commutation relations and that the resulting partition function correctly describes a system of fermions.

Given a generic quantum observable $\hat{A}$ its thermal average in the grand canonical ensemble is given by

$$\begin{aligned} \langle \hat{A} \rangle &= \frac{1}{\mathcal{Z}} \mathrm{Tr}[\hat{A}[\hat{\psi}_\sigma^+, \hat{\psi}_\sigma] \, e^{-\beta(\hat{H} - \mu\hat{N})}] \\ &= \frac{1}{\mathcal{Z}} \int \mathcal{D}[\bar{\psi}_\sigma, \psi_\sigma] \, A[\bar{\psi}_\sigma, \psi_\sigma] \, e^{-S_E[\bar{\psi}_\sigma, \psi_\sigma]} = \langle A \rangle \; , \quad (4.58) \end{aligned}$$

where $A[\bar{\psi}_\sigma, \psi_\sigma]$ is the Grassmann symbol associated to $\hat{A}[\hat{\psi}^+, \hat{\psi}]$. It is very important to stress again that this thermal average is expressed as a path integral over fermionic fields with anti-periodic boundary conditions. In general, the correct mapping from a quantum observable $\hat{A} = A[\hat{\psi}^+_\sigma, \hat{\psi}_\sigma]$ to its Grassmann symbol $A[\bar{\psi}_\sigma, \psi_\sigma]$ is obtained by replacing $\hat{\psi}_\sigma$ with ψ_σ and $\hat{\psi}^+_\sigma$ with $\bar{\psi}_\sigma$ without changing the operator ordering.

4.4.1 Non-Interacting Fermions in a Generic Trapping Potential

In the case of non-interacting two-spin-component fermions in a generic trapping potential $U(\mathbf{r})$, the Euclidean Lagrangian density is quite simple

$$\mathscr{L}_E = \sum_{\sigma=\uparrow,\downarrow} \bar{\psi}_\sigma \left[\hbar\partial_\tau - \frac{\hbar^2}{2m}\nabla^2 + U(\mathbf{r}) - \mu\right] \psi_\sigma \, . \tag{4.59}$$

It is then useful to consider the single-particle stationary Schrödinger problem

$$\left[-\frac{\hbar^2}{2m}\nabla^2 + U(\mathbf{r})\right] \phi_\alpha(\mathbf{r}) = \epsilon_\alpha \, \phi_\alpha(\mathbf{r}) \, , \tag{4.60}$$

where ϵ_α is the single-particle energy and $\phi_\alpha(\mathbf{r})$ the corresponding single-particle eigenfunction. As usual, α denotes the set of quantum numbers of the single quantum particle. Please remember that

$$\int d^3\mathbf{r} \, \phi^*_\alpha(\mathbf{r}) \, \psi_{\alpha'} = \delta_{\alpha,\alpha'} \tag{4.61}$$

with $\delta_{\alpha,\alpha'}$ the Kronecker delta. The Grassmann field can be then written as

$$\psi_\sigma(\mathbf{r}, t) = \sum_\alpha c_{\alpha,\sigma}(\tau) \, \phi_\alpha(\mathbf{r}) \, . \tag{4.62}$$

Inserting it into the Euclidean Lagrangian density and integrating over spatial coordinates, we obtain the following Euclidean action

$$S_E = \sum_{\alpha,\sigma} \int_0^{\hbar\beta} d\tau \, \bar{c}_{\alpha,\sigma}(\tau)(\hbar\frac{\partial}{\partial\tau} + \epsilon_\alpha - \mu)c_{\alpha,\sigma}(\tau) \, . \tag{4.63}$$

We further simplify the problem setting

$$c_{\alpha,\sigma}(\tau) = \sum_{n=-\infty}^{+\infty} c_{\alpha,n,\sigma}\, e^{-i\omega_n \tau} \tag{4.64}$$

with ω_n the fermionic Matsubara frequencies. In this way, keeping in mind that

$$\int_0^{\hbar\beta} d\tau\, e^{i(\omega_n - \omega_{n'})\tau} = \hbar\beta\, \delta_{n,n'} \ , \tag{4.65}$$

we find immediately

$$S_E = \hbar \sum_{\alpha} \sum_{n=-\infty}^{+\infty} \sum_{\sigma=\uparrow,\downarrow} \bar{c}_{\alpha,n,\sigma}[\beta(-i\hbar\omega_n + \epsilon_\alpha - \mu)]c_{\alpha,n,\sigma} \ . \tag{4.66}$$

Thus, the grand canonical partition function of a non-interacting gas of fermions with single-particle energies ϵ_α is given by

$$\begin{aligned} \mathcal{Z} &= \int \mathcal{D}[\bar{c}_{\alpha,n,\sigma}, c_{\alpha,n,\sigma}]\, e^{-S_E/\hbar} \\ &= \prod_{\alpha} \prod_{\sigma=\uparrow,\downarrow} \prod_{n=-\infty}^{+\infty} [-i\pi(2n+1) + \beta(\epsilon_\alpha - \mu)] \\ &= \prod_{\alpha} \prod_{\sigma=\uparrow,\downarrow} (1 + e^{\beta(\epsilon_\alpha - \mu)}) \ , \end{aligned} \tag{4.67}$$

and the grand potential follows as

$$\Omega = -\frac{1}{\beta} \ln(\mathcal{Z}) = -2k_B T \sum_{\alpha} \ln\left(1 + e^{\beta(\epsilon_\alpha - \mu)}\right) \ , \tag{4.68}$$

where the factor 2 comes from the two possible spin states.

4.4.2 Ideal and Uniform Fermi Gas

We analyze a D-dimensional ($D = 1, 2, 3$) Fermi gas of noninteracting fermions in a D-dimensional box of volume L^D. Under these conditions exact analytical results can be obtained in any spatial dimension D. In this case the Euclidean Lagrangian density is given by

$$\mathscr{L}_E = \sum_{\sigma=\uparrow,\downarrow} \bar{\psi}_\sigma \left[\hbar\partial_\tau - \frac{\hbar^2}{2m}\nabla^2 - \mu \right] \psi_\sigma \, . \tag{4.69}$$

We use the Fourier decomposition

$$\psi_\sigma(\mathbf{r},\tau) = \frac{1}{\sqrt{L^D}} \sum_{n=-\infty}^{+\infty} \sum_{\mathbf{k}} c_{\mathbf{k},n,\sigma} \; e^{i(\mathbf{k}\cdot\mathbf{r}-\omega_n\tau)} \; , \tag{4.70}$$

where $c_{\mathbf{k},n,\sigma}$ are Grassmann variables with $\mathbf{k}$ the D-dimensional linear wavevector and $\omega_n = \pi(2n+1)/(\beta\hbar)$ the fermionic Matsubara frequencies. Here $c_{\mathbf{k},n,\sigma}$ plays the role of the Fourier transform of $\psi_\sigma(\mathbf{r},\tau)$, and the quantization of frequencies ω_n is a consequence of the anti-periodicity $\psi_\sigma(\mathbf{r},\tau+\hbar\beta) = -\psi_\sigma(\mathbf{r},\tau)$. It follows that the Euclidean action can be written in a diagonal form as

$$S_E[\bar{\psi}_\sigma, \psi_\sigma] = \sum_{\sigma=\uparrow,\downarrow} \sum_{n=-\infty}^{+\infty} \sum_{\mathbf{k}} \hbar\lambda_{\mathbf{k},n} \, \bar{c}_{\mathbf{k},n,\sigma} c_{\mathbf{k},n,\sigma} \tag{4.71}$$

where

$$\lambda_{\mathbf{k},n} = \beta\,(-i\hbar\omega_n + \frac{\hbar^2 k^2}{2m} - \mu) \, . \tag{4.72}$$

Taking into account that

$$\begin{aligned} \mathcal{Z} &= \int \mathcal{D}[\bar{c}_{\mathbf{k},n,\sigma}, c_{\mathbf{k},n,\sigma}] \exp\left\{ - \sum_{\sigma=\uparrow,\downarrow} \sum_{n=-\infty}^{+\infty} \sum_{\mathbf{k}} \lambda_{\mathbf{k},n} \bar{c}_{\mathbf{k},n,\sigma} c_{\mathbf{k},n,\sigma} \right\} \\ &= \prod_{\sigma=\uparrow,\downarrow} \prod_{n=-\infty}^{+\infty} \prod_{\mathbf{k}} \lambda_{\mathbf{k},n} \, , \end{aligned} \tag{4.73}$$

and also Eq. (4.34) of the fermionic Matsubara frequencies, the grand potential reads

$$\Omega = -\frac{1}{\beta} \ln \mathcal{Z} = -\frac{2}{\beta} \sum_{\mathbf{k}} \ln\left(1 + e^{-\beta\xi_k}\right) \tag{4.74}$$

with

$$\xi_k = \frac{\hbar^2 k^2}{2m} - \mu \tag{4.75}$$

the shifted single-particle kinetic energy. The factor 2 in the fermionic grand potential is due to the two possible spin states.

In the continuum limit the number density reads

$$n = -\frac{1}{L^D}\frac{\partial \Omega}{\partial \mu} = 2\int \frac{d^D\mathbf{k}}{(2\pi)^D}\frac{1}{e^{\beta\xi_k}+1} \,. \tag{4.76}$$

At zero temperature this formula becomes

$$n = 2\int \frac{d^D\mathbf{k}}{(2\pi)^D}\Theta(-\xi_k) \,, \tag{4.77}$$

where $\Theta(x)$ is the Heaviside step function. After integration one obtains

$$\mu = \begin{cases} \frac{\hbar^2}{2m}(3\pi^2 n)^{2/3} & \text{for } D=3 \\ \frac{\hbar^2\pi}{m} n & \text{for } D=2 \\ \frac{\hbar^2\pi^2}{8m} n^2 & \text{for } D=1 \end{cases} \tag{4.78}$$

where the chemical potential μ can be identified with the Fermi energy ϵ_F. The zero-temperature internal energy E is immediately derived from the formula

$$\frac{E}{L^D} = \int_0^n dn\, \mu(n) \,, \tag{4.79}$$

from which we get

$$\frac{E}{L^D} = \begin{cases} \frac{3}{5}\frac{\hbar^2}{2m}(3\pi^2)^{2/3} n^{5/3} & \text{for } D=3 \\ \frac{1}{2}\frac{\hbar^2\pi}{m} n^2 & \text{for } D=2 \\ \frac{1}{3}\frac{\hbar^2\pi^2}{8m} n^3 & \text{for } D=1 \end{cases} \tag{4.80}$$

4.5 Repulsive Fermions: Hartree-Fock Approximation

We now investigate identical fermions with a repulsive contact interaction. This system is described by the Lagrangian density (4.56) with $g > 0$. For simplcity, we suppose that the system has the same number of spin up and spin down fermions, i.e., $n_\uparrow = n_\downarrow = n/2$, with n the total number density of fermions.

The problem cannot be solved exactly, and we adopt a mean-field Hartree-Fock approximation for the quartic term. In particular, we use this expression

$$g\,\bar{\psi}_\uparrow\,\bar{\psi}_\downarrow\,\psi_\downarrow\,\psi_\uparrow \simeq g\,\frac{n}{2}\bar{\psi}_\uparrow\psi_\uparrow + g\,\frac{n}{2}\bar{\psi}_\downarrow\psi_\downarrow - g\frac{n^2}{4}\,, \tag{4.81}$$

where

$$\frac{n}{2} = \langle\bar{\psi}_\uparrow\psi_\uparrow\rangle = \langle\bar{\psi}_\downarrow\psi_\downarrow\rangle\,. \tag{4.82}$$

In this way the fermionic Lagrangian density (4.56) becomes quadratic and given by

$$\mathscr{L}_{HF} = \sum_{\sigma=\uparrow,\downarrow} \bar{\psi}_\sigma \left[\hbar\partial_\tau - \frac{\hbar^2}{2m}\nabla^2 - \mu + \frac{1}{2}gn\right]\psi_\sigma - g\frac{n^2}{4}\,, \tag{4.83}$$

Having this Hartree-Fock quadratic Lagrangian density, we can perform the fermionic path integral calculation. It is important to observe that we have introduced explicitly the number density n in the Hartree-Fock Lagrangian density, which also contains the chemical potential μ. In the framework of the grand canonical ensemble, it means that n must be considered as a variational parameter that is self-consistently determined by minimizing the grand potential Ω.

4.5.1 Justification of the Mean-Field Treatment

The justification of the mean-field approximation of Eq. (4.81) follows from the fact that, given two Grassmann symbols A and B, associated to the fermionic operators $\hat{A}$ and $\hat{B}$, with averages $\langle A\rangle = \langle\hat{A}\rangle$ and $\langle B\rangle = \langle\hat{B}\rangle$, the mean-field result holds

$$A\,B \simeq \langle A\rangle\,B + A\,\langle B\rangle - \langle A\rangle\,\langle B\rangle \tag{4.84}$$

neglecting quadratic fluctuations. To prove Eq. (4.84) we set

$$A = \langle A\rangle + \delta_A \tag{4.85}$$

$$B = \langle B\rangle + \delta_B \tag{4.86}$$

where

$$\begin{aligned}\delta_A &= A - \langle A \rangle && (4.87)\\ \delta_B &= B - \langle B \rangle && (4.88)\end{aligned}$$

are the fluctuations of A and B. Then we write

$$A\,B = \Big(\langle A\rangle + \delta_A\Big)\Big(\langle B\rangle + \delta_B\Big) = \langle A\rangle\,\langle B\rangle + \langle A\rangle\,\delta_B + \langle B\rangle\,\delta_A + \delta_A\,\delta_B\ . \quad (4.89)$$

Neglecting the term $\delta_A\,\delta_B$ and taking into account the definition of fluctuations, the previous equation gives

$$\begin{aligned}A\,B &\simeq \langle A\rangle\,\langle B\rangle + \langle A\rangle\,\delta_B + \langle B\rangle\,\delta_A\\ &= \langle A\rangle\,\langle B\rangle + \langle A\rangle\,(B - \langle B\rangle) + \langle B\rangle\,(A - \langle A\rangle)\\ &= \langle A\rangle\,\langle B\rangle + \langle A\rangle\,B - \langle A\rangle\,\langle B\rangle + \langle B\rangle\,A - \langle B\rangle\,\langle A\rangle\\ &= \langle A\rangle\,B + A\,\langle B\rangle - \langle A\rangle\,\langle B\rangle\ , && (4.90)\end{aligned}$$

by using the fact that averages commute with themselves and also with Grassmann symbols.

4.5.2 Grand Potential

Taking into account Eq. (4.81) and repeating the procedure of the previous section, it is straightforward to show that the mean-field partition function is given by

$$\begin{aligned}\mathcal{Z}_{MF} &= e^{\beta g \frac{n^2}{4} L^D} \prod_{\sigma=\uparrow,\downarrow} \prod_{\mathbf{k}} \prod_{n=-\infty}^{+\infty} \left(-i\pi(2n+1) + \beta E_k\right)\\ &= e^{\beta g \frac{n^2}{4} L^D} \prod_{\sigma=\uparrow,\downarrow} \prod_{\mathbf{k}} (1 + e^{-\beta E_k}) && (4.91)\end{aligned}$$

and the mean-field grand potential reads

$$\frac{\Omega_{MF}}{L^D} = -\frac{2}{\beta} \int \frac{d^D\mathbf{k}}{(2\pi)^D} \ln\left(1 + e^{-\beta E_k}\right) - g\,\frac{n^2}{4}\ , \quad (4.92)$$

where the single-particle Hartree-Fock energy E_k given by

$$E_k = \frac{\hbar^2 k^2}{2m} - \mu + \frac{1}{2} g\, n\ . \quad (4.93)$$

As previously discussed, the number density n which appears in the grand potential Ω is a variational parameter that is determined from the equation

$$\frac{\partial \Omega_{MF}}{\partial n} = 0 \,. \tag{4.94}$$

It gives

$$n = 2 \int \frac{d^D\mathbf{k}}{(2\pi)^D} \frac{1}{e^{\beta E_k} + 1} \,, \tag{4.95}$$

and, at zero-temperature, after integration one obtains

$$\mu = \begin{cases} \frac{\hbar^2}{2m}(3\pi^2 n)^{2/3} + \frac{1}{2} g\, n & \text{for } D = 3 \\ \frac{\hbar^2\pi}{m} n + \frac{1}{2} g\, n & \text{for } D = 2 \\ \frac{\hbar^2\pi^2}{8m} n^2 + \frac{1}{2} g\, n & \text{for } D = 1 \end{cases} \tag{4.96}$$

In the three-dimensional case, this result was obtained for the first time in 1957 by Kerson Huang and Chen Ning Yang. Also in this case the zero-temperature internal energy E is easily derived by integrating the chemical potential μ with respect to the local density n, finding

$$\frac{E}{L^D} = \begin{cases} \frac{3}{5}\frac{\hbar^2}{2m}(3\pi^2)^{2/3} n^{5/3} + \frac{1}{4}\, g\, n^2 & \text{for } D = 3 \\ \frac{1}{2}\frac{\hbar^2\pi}{m} n^2 + \frac{1}{4}\, g\, n^2 & \text{for } D = 2 \\ \frac{1}{3}\frac{\hbar^2\pi^2}{8m} n^3 + \frac{1}{4}\, g\, n^2 & \text{for } D = 1 \end{cases} \tag{4.97}$$

4.5.3 Stoner Instability

Let us consider a two-component three-dimensional ($D = 3$) Fermi gas with repulsive contact interaction of strength g. Within the mean-field Hartree-Fock approximation, we now find the critical strength g_c above which the zero-temperature uniform unpolarized configuration becomes unstable (Stoner instability).

Assuming that $\mu_\uparrow \neq \mu_\downarrow$ and $n_\uparrow \neq n_\downarrow$, the mean-field Hartree-Fock approximation for the quartic term gives

$$g\, \bar{\psi}_\uparrow \bar{\psi}_\downarrow \psi_\downarrow \psi_\uparrow \simeq g\, n_\downarrow \bar{\psi}_\uparrow \psi_\uparrow + g\, n_\uparrow \bar{\psi}_\downarrow \psi_\downarrow - g n_\uparrow n_\downarrow \,. \tag{4.98}$$

In this way the effective Lagrangian density is given by

$$\mathscr{L}_{HF} = \sum_{\sigma=\uparrow,\downarrow} \bar{\psi}_\sigma \left[\hbar\partial_\tau - \frac{\hbar^2}{2m}\nabla^2 - \mu_\sigma + g n_{-\sigma} \right] \psi_\sigma - g n_\uparrow n_\downarrow \,, \tag{4.99}$$

and the mean-field grand potential reads

$$\frac{\Omega_{MF}}{L^3} = -\frac{1}{\beta} \sum_{\sigma=\uparrow,\downarrow} \int \frac{d^3\mathbf{k}}{(2\pi)^3} \ln\left(1 + e^{-\beta E_k^{(\sigma)}}\right) - g n_\uparrow n_\downarrow \,, \tag{4.100}$$

where the single-particle Hartree-Fock energy $E_k^{(\sigma)}$ given by

$$E_k^{(\sigma)} = \frac{\hbar^2 k^2}{2m} - \mu_\sigma + g n_{\sigma'} \tag{4.101}$$

and $\sigma' \neq \sigma$. At zero-temperature, after integration over momenta one finds the following internal energy density

$$\frac{E}{L^3} = \frac{\Omega}{L^3} + \mu_\uparrow n_\uparrow + \mu_\downarrow n_\downarrow = \frac{3}{5}\frac{\hbar^2}{2m}(6\pi^2)^{2/3}\left(n_\uparrow^{5/3} + n_\downarrow^{5/3}\right) + g n_\uparrow n_\downarrow \,, \tag{4.102}$$

and chemical potentials

$$\mu_\sigma = \frac{\hbar^2}{2m}(6\pi^2)^{2/3} n_\sigma^{2/3} + g n_{\sigma'} \tag{4.103}$$

with $\sigma \neq \sigma'$. Given the Jacobian

$$J = \begin{pmatrix} \frac{\partial \mu_\uparrow}{\partial n_\uparrow} & \frac{\partial \mu_\uparrow}{\partial n_\downarrow} \\ \frac{\partial \mu_\downarrow}{\partial n_\uparrow} & \frac{\partial \mu_\downarrow}{\partial n_\downarrow} \end{pmatrix} = \begin{pmatrix} \frac{2}{3}\frac{\hbar^2}{2m}(6\pi^2)^{2/3} n_\uparrow^{-1/3} & g \\ g & \frac{2}{3}\frac{\hbar^2}{2m}(6\pi^2)^{2/3} n_\downarrow^{-1/3} \end{pmatrix} \tag{4.104}$$

the critical strength g_c of the unpolarized configuration is obtained by setting $n_\uparrow = n_\downarrow = n/2$ and the determinant of J equal to zero. Notice that this Jacobian J is nothing else than the Hessian of the second-partial derivatives of the energy E with respect to the densities n_σ. It follows

$$g_c = \frac{2}{3}\frac{\hbar^2}{2m}(6\pi^2)^{2/3}\left(\frac{n}{2}\right)^{-1/3} . \tag{4.105}$$

At fixed number density n, for $g > g_c$ the determinant of the Jacobian J is negative, and this implies the uniform configuration is energetically unstable. On the contrary, for $0 \leq g < g_c$ the determinant of J is positive and the uniform configuration is stable.

4.6 Attractive Fermions

In this section we investigate identical fermions with an attractive contact interaction in three spatial dimensions. This system is described by the Lagrangian density (4.56) with $g < 0$. In addition, we suppose that the system has the same number of spin-up and spin-down fermions, i.e., $n_\uparrow = n_\downarrow = n/2$, with n the total number density of fermions. As discussed in the previous section, the problem cannot be solved exactly due to the presence of the quartic term in the Lagrangian density.

4.6.1 Mean-Field BCS Treatment

Also in this case we adopt a mean-field treatment which is, however, quite different with respect to the Hartree-Fock one of repulsive fermions. We suppose that, due to the attractive interaction, fermions with opposite spin can create a sort of bosonic particle with spin zero. In particular, considering again the quartic term of the Euler Lagrangian density, we use here a different mean-field approximation with respect to the Hartree-Fock one adopted for repulsive fermions. We write

$$-g_0\, \hat{\psi}_\uparrow^+\, \hat{\psi}_\downarrow^+\, \hat{\psi}_\downarrow\, \hat{\psi}_\uparrow \simeq -\Delta_0^*\, \hat{\psi}_\downarrow \hat{\psi}_\uparrow - \Delta_0 \hat{\psi}_\uparrow^+ \hat{\psi}_\downarrow^+ + \frac{|\Delta_0|^2}{g_0}\,, \qquad (4.106)$$

where $g_0 = -g > 0$ and

$$\Delta_0 = g_0\, \langle \hat{\psi}_\downarrow\, \hat{\psi}_\uparrow \rangle \qquad (4.107)$$

is a complex order parameter usually called the pairing energy or the pairing energy gap. For attractive fermions, this Bardeen-Cooper-Schrieffer (BCS) approximation gives a much better agreement with exact calculations (and also with experimental data) than the Hartree-Fock approximation. In this way the corresponding Lagrangian density (4.56) of the fermionic path integral formulation becomes quadratic and is given by

$$\mathscr{L}_{BCS} = \sum_{\sigma=\uparrow,\downarrow} \bar{\psi}_\sigma \left[\hbar\partial_\tau - \frac{\hbar^2}{2m}\nabla^2 - \mu\right] \psi_\sigma - \Delta_0\, \psi_\downarrow\, \psi_\uparrow - \Delta_0 \bar{\psi}_\uparrow\, \bar{\psi}_\downarrow + \frac{\Delta_0^2}{g_0}\,. \qquad (4.108)$$

In this mean-field treatment, for the sake of simplicity, we will assume that the pairing gap Δ_0 is real and constant both in space and time. Introducing the Nambu spinor

$$\Psi(\mathbf{r},\tau) = \begin{pmatrix} \psi_\uparrow(\mathbf{r},\tau) \\ \bar{\psi}_\downarrow(\mathbf{r},\tau) \end{pmatrix} \tag{4.109}$$

the Lagrangian density can be rewritten as

$$\mathscr{L}_{BCS} = \frac{\Delta_0^2}{g_0} + \bar{\Psi}^T \hat{M} \Psi \ , \tag{4.110}$$

where

$$\hat{M} = \begin{pmatrix} \hbar\partial_\tau - \frac{\hbar^2}{2m}\nabla^2 - \mu & -\Delta_0 \\ -\Delta_0 & \hbar\partial_\tau + \frac{\hbar^2}{2m}\nabla^2 + \mu \end{pmatrix} \tag{4.111}$$

is a 2×2 matrix of differential operators. By using the Fourier expansion

$$\psi_\sigma(\mathbf{r},\tau) = \frac{1}{L^{3/2}} \sum_{n=-\infty}^{+\infty} \sum_{\mathbf{k}} c_{\mathbf{k},n,\sigma} \ e^{i(\mathbf{k}\cdot\mathbf{r} - \omega_n \tau)} \ , \tag{4.112}$$

the Nambu spinor in Fourier space

$$\Psi_{\mathbf{k},n} = \begin{pmatrix} c_{\mathbf{k},n,\uparrow} \\ \bar{c}_{-\mathbf{k},-n,\downarrow} \end{pmatrix} \ , \tag{4.113}$$

and the orthonormalization

$$\int_0^{\hbar\beta} d\tau \int_{L^3} d^3\mathbf{r} \, e^{i[(\mathbf{k}-\mathbf{k}')\cdot\mathbf{r} - (\omega_n - \omega_{n'})\tau]} = \hbar\beta \, L^3 \, \delta_{\mathbf{k},\mathbf{k}'} \, \delta_{n,n'} \ , \tag{4.114}$$

we obtain the following BCS action

$$S_{BCS} = \hbar\beta L^3 \frac{\Delta_0^2}{g_0} + \hbar \sum_{n=-\infty}^{+\infty} \sum_{\mathbf{k}} \bar{\Psi}_{\mathbf{k},n}^T \hat{M}_{\mathbf{k},n} \Psi_{\mathbf{k},n} \ , \tag{4.115}$$

where

$$\hat{M}_{\mathbf{k},n} = \beta \begin{pmatrix} -i\hbar\omega_n + \frac{\hbar^2 k^2}{2m} - \mu & -\Delta_0 \\ -\Delta_0 & -i\hbar\omega_n - \frac{\hbar^2 k^2}{2m} + \mu \end{pmatrix} \ . \tag{4.116}$$

4.6.2 BCS Elementary Excitations

The BCS action (4.115) can be written in a diagonal form by performing a fermionic Bogoliubov transformation

$$\begin{pmatrix} f_{\mathbf{k},n,\uparrow} \\ \bar{f}_{\mathbf{k},n,\uparrow} \end{pmatrix} = \begin{pmatrix} u_{\mathbf{k}} & -v_{\mathbf{k}} \\ v_{\mathbf{k}} & u_{\mathbf{k}} \end{pmatrix} \begin{pmatrix} c_{\mathbf{k},n,\uparrow} \\ \bar{c}_{-\mathbf{k},-n,\downarrow} \end{pmatrix} . \tag{4.117}$$

In particular, imposing that

$$u_{\mathbf{k}}^2 = 1 - v_{\mathbf{k}}^2 = \frac{1}{2}\left(1 + \frac{\frac{\hbar^2 k^2}{2m} - \mu}{E_k}\right) \tag{4.118}$$

with

$$E_k = \sqrt{\left(\frac{\hbar^2 k^2}{2m} - \mu\right)^2 + \Delta_0^2} \tag{4.119}$$

the single-particle BCS elementary excitations, one finds

$$\begin{aligned} S_{BCS} &= \frac{\Delta_0^2}{g_0} L^3 \hbar\beta \\ &+ \hbar \sum_{\sigma=\uparrow,\downarrow} \sum_{n=-\infty}^{+\infty} \sum_{\mathbf{k}} (\bar{f}_{\mathbf{k},n,\uparrow}, f_{-\mathbf{k},-n,\downarrow}) \\ &\times \begin{pmatrix} \beta(-i\hbar\omega_n + E_k) & 0 \\ 0 & \beta(-i\hbar\omega_n - E_k) \end{pmatrix} \begin{pmatrix} f_{\mathbf{k},n,\uparrow} \\ \bar{f}_{-\mathbf{k},-n,\downarrow} \end{pmatrix} \\ &= \frac{\Delta_0^2}{g_0} L^3 \hbar\beta + \hbar \sum_{\sigma=\uparrow,\downarrow} \sum_{n=-\infty}^{+\infty} \sum_{\mathbf{k}} \lambda_{\mathbf{k},n}\, \bar{f}_{\mathbf{k},n,\sigma} f_{\mathbf{k},n,\sigma} \,, \end{aligned} \tag{4.120}$$

where

$$\lambda_{\mathbf{k},n} = \beta\left(-i\hbar\omega_n + E_k\right) . \tag{4.121}$$

Notice that we have used the formal property $\omega_{-n} = \omega_n$ with the general definition of fermionic Matsubara frequencies $\omega_n = \pi n/(\hbar\beta)$ with odd n. In this fermionic problem, the 2×2 Bogoliubov matrix of Eq. (4.117) is unitary, and the eigenvalues of the matrix $\hat{M}_{\mathbf{k},n}$ of Eq. (4.116) are the two elements of the diagonal matrix of Eq. (4.120).

In the BCS framework, BCS excitations E_k are positive definite. The BCS quasiparticles are a mixture of fermionic particles and holes, and their energy spectrum is constructed to account for this duality. When $\Delta = 0$ one gets $E_k = |\hbar^2 k^2/(2m) - \mu|$, and the modulus reflects that the energy of the excitation depends on whether the state is above or below the chemical potential μ. In the BCS context, an excitation always requires positive energy, even if it involves removing a particle from below the Fermi surface.

Having the diagonal BCS action, the mean-field partition function is easily derived:

$$\begin{aligned} \mathcal{Z}_{MF} &= \int \mathcal{D}[\bar{f}_{\mathbf{k},n,\sigma}, f_{\mathbf{k},n,\sigma}]\, e^{-S_E[\bar{f}_{\mathbf{k},n,\sigma}, f_{\mathbf{k},n,\sigma}]/\hbar} \\ &= e^{-\beta \frac{\Delta_0^2}{g_0} L^3} \prod_{\sigma=\uparrow,\downarrow} \prod_{\mathbf{k}} \prod_{n=-\infty}^{+\infty} \lambda_{\mathbf{k},n} = e^{-\beta \frac{\Delta_0^2}{g_0} L^3} \prod_{\sigma=\uparrow,\downarrow} \prod_{\mathbf{k}} (1 + e^{\beta E_k}) \end{aligned} \tag{4.122}$$

and the mean-field grand potential can then be written as

$$\Omega_{MF} = -\frac{1}{\beta} \ln(\mathcal{Z}_{MF}) = \Omega^{(0)} + \Omega^{(T)} \tag{4.123}$$

including the mean-field pairing contribution

$$\Omega_0 = \frac{\Delta_0^2}{g_0} L^3 \tag{4.124}$$

and the grand potential of the fermionic single-particle excitations

$$\Omega^{(T)} = -\frac{2}{\beta} \sum_{\mathbf{k}} \ln\left(1 + e^{-\beta E_k}\right) . \tag{4.125}$$

4.6.3 Zero-Point Energy of BCS Excitations

We have not yet considered the zero-point energy of Bogoliubov fermionic fields. The zero-point energy of the BCS elementary excitations reads

$$\Omega_g^{(0)} = -\frac{1}{2} \sum_{\sigma=\uparrow,\downarrow} \sum_{\mathbf{k}} (E_k - \xi_k) = -\sum_{\mathbf{k}} (E_k - \xi_k) \tag{4.126}$$

with $E_{\mathbf{k}}$ given by Eq. (4.119) and $\xi_k = \hbar^2 k^2/(2m) - \mu$. The term (4.126) pops out because one has interchanged the Bogoliubov

Grassmann fields $f_{\mathbf{k},n,\sigma}$ and $\bar{f}_{\mathbf{k},n,\sigma}$. One must do it in the discretized version of the path integral, and only at the end the continuum time limit must be taken, with the appearance of a convergence factor $e^{i\omega_n 0^+}$ which produces the zero-point energy of the interacting field. Notice that for non-interacting fermions, where there are particle excitations above the Fermi surface ($\xi_k > 0$) and hole excitations below the Fermi surface ($\xi_k < 0$), one has $E_k = \xi_k$ and $\Omega_g^{(0)} = 0$. Alternatively, one can deduce the zero-point energy starting from the continuum version of the path integral and performing an evaluation without the convergence factor of the infinite product of fermionic Matsubara frequencies.

In the next subsections we will see that the zero-point energy of BCS excitations is crucial to get reliable results at low temperatures, despite the fact that the zero-point energy of the fermionic field is divergent. The divergence of the zero-point BCS energy can be regularized by adopting different regularization schemes (for dimensional regularization, see Papenbrock and Bertsch (1999)). In the next sections we use Eq. (4.126) and the cutoff regularization because it is physically sound in the context of metallic superconductors.

4.6.4 Gap and Number Equations

We add the zero-point energy term $\Omega_g^{(0)}$ contribution to the grand potential Ω_{MF} obtaining

$$\Omega = \Omega_{MF} + \Omega_g^{(0)} = \frac{\Delta_0^2}{g_0} L^3 - \sum_{\mathbf{k}} \left[\frac{2}{\beta} \ln\left(1 + e^{-\beta E_k}\right) + (E_k - \xi_k) \right] . \tag{4.127}$$

The energy gap Δ_0 which appears in the fermionic energy spectrum (4.119) is determined by imposing that

$$\left(\frac{\partial \Omega}{\partial \Delta_0} \right)_{\mu, T, L^3} = 0 \, , \tag{4.128}$$

from which one gets

$$\frac{1}{g_0} = \frac{1}{L^3} \sum_{\mathbf{k}} \frac{\tanh\left(\frac{\beta E_k}{2}\right)}{2 E_k} \tag{4.129}$$

that is the so-called gap equation with $\tanh(x/2) = 1-(2/(e^x+1))$. At zero temperature $T = 0$, i.e., for $\beta \to \infty$ where $\tanh(\beta E_k/2) \to 0$, the gap equation becomes

$$\frac{1}{g_0} = \frac{1}{L^3}\sum_{\mathbf{k}} \frac{1}{2E_k} \,. \tag{4.130}$$

The number density n is instead obtained from

$$n = -\frac{1}{L^3}\left(\frac{\partial \Omega}{\partial \mu}\right)_{T,L^3} \,. \tag{4.131}$$

Observing that

$$\begin{aligned}\left(\frac{\partial \Omega}{\partial \mu}\right)_{T,L^3} &= \left(\frac{\partial \Omega}{\partial \mu}\right)_{T,L^3,\Delta_0} + \left(\frac{\partial \Omega}{\partial \Delta_0}\right)_{\mu,T,L^3}\left(\frac{\partial \Delta_0}{\partial \mu}\right)_{T,L^3} \\ &= \left(\frac{\partial \Omega}{\partial \mu}\right)_{T,L^3,\Delta_0}\end{aligned} \tag{4.132}$$

we find the so-called number equation

$$n = \frac{1}{L^3}\sum_{\mathbf{k}}\left[1 - \frac{\xi_k}{E_k}\tanh\left(\frac{\beta E_k}{2}\right)\right] \,. \tag{4.133}$$

At zero temperature it becomes

$$n = \frac{1}{L^3}\sum_{\mathbf{k}}\left[1 - \frac{\xi_k}{E_k}\right] \,. \tag{4.134}$$

Moreover, if the energy gap is zero, i.e., with $\Delta_0 = 0$, this last equation gives

$$n = \frac{1}{L^3}\sum_{\mathbf{k}}\left[1 - \frac{\xi_k}{|\xi_k|}\right] = \frac{2}{L^3}\sum_{\mathbf{k}}\Theta\left(\mu - \frac{\hbar^2 k^2}{2m}\right) \,, \tag{4.135}$$

that is the familiar number density of non-interacting fermions with two spin components. Here, we have used $1 - \xi_k/|\xi_k| = 2\Theta(-\xi)$ with $\Theta(x)$ the Heaviside step function. In three-dimensions

($D = 3$) the gap equation (4.129) is ultraviolet divergent. It becomes convergent by introducing an ultraviolet cutoff Λ such that

$$\frac{1}{g_0} = \frac{1}{L^3} \sum_{|\mathbf{k}|<\Lambda} \frac{\tanh\left(\frac{\beta E_k}{2}\right)}{2E_k} \ . \tag{4.136}$$

In the BCS theory of metallic superconductors, phonons produce the attractive interaction between conduction electrons. Phonons can transfer at most the energy $\hbar\omega_D$ to electrons, where ω_D is the Debye frequency. Consequently, the ultraviolet cutoff Λ is such that $\hbar^2\Lambda^2/(2m) = \mu + \hbar\omega_D$. For this specific physical system, there is also the infrared cutoff k_m with energy $\hbar^2 k_m^2/(2m) = \mu - \hbar\omega_D$. In other words,

$$\frac{1}{g_0} = \frac{1}{L^3} \sum_{k_m<|\mathbf{k}|<\Lambda} \frac{\tanh\left(\frac{\beta E_k}{2}\right)}{2E_k} \tag{4.137}$$

and there is Cooper pairing only near the Fermi surface $|\mathbf{k}| = k_F = \sqrt{2m\mu/\hbar^2}$. Thus, one models the attractive electron-electron potential mediated by phonons as follows

$$V(\mathbf{r}, \mathbf{r}') = \begin{cases} -g_0\, \delta^{(3)}(\mathbf{r} - \mathbf{r}')\,(1 - \delta_{\sigma,\sigma'}) & \text{for } |\xi_k|, |\xi_{k'}| < \hbar\omega_D \\ 0 & \text{otherwise} \end{cases} , \tag{4.138}$$

where $\mathbf{k}$ and $\mathbf{k}'$ are the wavenumbers of the two electrons. In reciprocal Fourier space this means

$$V_{\mathbf{k},\mathbf{k}'} = \begin{cases} -g_0\,(1 - \delta_{\sigma,\sigma'}) & \text{for } |\xi_k|, |\xi_{k'}| < \hbar\omega_D \\ 0 & \text{otherwise} \end{cases} \tag{4.139}$$

and also

$$\Delta_{\mathbf{k},\mathbf{k}'} = \begin{cases} \Delta_0(1 - \delta_{\sigma,\sigma'}) & \text{for } |\xi_k|, |\xi_{k'}| < \hbar\omega_D \\ 0 & \text{otherwise} \end{cases} \tag{4.140}$$

for the pairing energy gap. Strictly speaking, it is not necessary to include the constraint $(1 - \delta_{\sigma,\sigma'})$. In fact, in the case of contact interaction, two fermions can interact only if they have opposite spin: the total wavefunction of the two-particle problem must be antisymmetric. Since the spatial wavefunction is symmetric due to the contact interaction (both fermions are at the same point in space), the spin wavefunction must be antisymmetric to preserve the overall antisymmetry of the total wavefunction. There are many subtleties in the mathematical derivation of the attractive interaction between electrons starting from the many-body problem of electrons moving in an ionic lattice. Without going into details, one can say that as an electron moves through the metal, it distorts the lattice, creating a positive charge density that attracts another electron. This leads to an effective attractive force between electrons despite their intrinsic Coulomb repulsion.

In the thermodynamic limit, where the volume L^3 goes to infinity, $\sum_{\mathbf{k}}$ can be replaced by $L^3 \int d^3\mathbf{k}/(2\pi)^3$ and the gap equation becomes

$$\begin{aligned} \frac{1}{g_0} &= \frac{1}{2\pi^2} \int_{\mu-\hbar\omega_D}^{\mu+\hbar\omega_D} dk\, k^2 \, \frac{\tanh\left(\frac{\beta E_k}{2}\right)}{2E_k} \\ &= \int_{-\hbar\omega_D}^{\hbar\omega_D} d\xi\, \mathscr{D}(\xi)\, \frac{\tanh\left(\frac{\beta}{2}\sqrt{\xi^2+\Delta_0^2}\right)}{2\sqrt{\xi^2+\Delta_0^2}} \,, \end{aligned} \tag{4.141}$$

where

$$\begin{aligned} \mathscr{D}(\xi) &= \frac{1}{2\pi^2} \int_0^{+\infty} dk\; k^2\, \delta\left(\xi - \left(\frac{\hbar^2 k^2}{2m} - \mu\right)\right) \\ &= \frac{1}{4\pi^2} \left(\frac{2m}{\hbar^2}\right)^{3/2} (\mu+\xi)^{1/2} \end{aligned} \tag{4.142}$$

is the density of states. In metals the condition $\hbar\omega_D \ll \mu$ is always satisfied, consequently, because $\xi \in [-\hbar\omega_D, \hbar\omega_D]$, we use the approximation $\mathscr{D}(\xi) \simeq \mathscr{D}(0)$ with

$$\mathscr{D}(0) = \frac{1}{4\pi^2} \left(\frac{2m}{\hbar^2}\right)^{3/2} \mu^{1/2} \tag{4.143}$$

the density of states at the Fermi surface. In this way the gap equation (4.136) becomes

$$\frac{1}{g_0 \mathscr{D}(0)} = \int_0^{\hbar\omega_D} \frac{\tanh\left(\frac{\beta}{2}\sqrt{\xi^2+\Delta_0^2}\right)}{\sqrt{\xi^2+\Delta_0^2}}\, d\xi \,. \tag{4.144}$$

It is useful a remark: in the two-dimensional case, the density of states $\mathscr{D}(\xi)$ is constant, i.e., the formula $\mathscr{D}(\xi) = \mathscr{D}(0)$ is exact.

At zero-temperature ($\beta \to +\infty$), from Eq. (4.144) with $\tanh(\beta E_k/2) \to 1$, we find a formula for the zero-temperature energy gap $\Delta_0(0)$:

$$\frac{1}{g_0\mathscr{D}(0)} = \int_0^{\hbar\omega_D} \frac{d\xi}{\sqrt{\xi^2 + \Delta_0(0)^2}} = \ln\left(\frac{\hbar\omega_D}{\Delta_0(0)} + \sqrt{1 + \frac{\hbar^2\omega_D^2}{\Delta_0(0)^2}}\right)$$
$$\simeq \ln\left(\frac{2\hbar\omega_D}{\Delta_0(0)}\right) = -\ln\left(\frac{\Delta_0(0)}{2\hbar\omega_D}\right) \tag{4.145}$$

under the condition $\Delta_0(0) \ll \hbar\omega_D$. Thus, we obtain the weak-coupling BCS result

$$\Delta_0(0) = 2\hbar\omega_D \exp\left(-\frac{1}{g_0\mathscr{D}(0)}\right) \tag{4.146}$$

for the zero-temperature energy gap, which shows the exponentially small contribution of the energy gap Δ_0 in the BCS theory. Because Δ_0 is extremely small, in this BCS regime one can safely used Eq. (4.135) finding

$$\mu = \frac{\hbar^2}{2m}(3\pi^2)^{2/3}n^{2/3} = \epsilon_F , \tag{4.147}$$

that is the Fermi energy of non-interacting fermions, and also, from Eq. (4.143), the density of states

$$\mathscr{D}(0) = \frac{m}{2\pi^2\hbar^2}(3\pi^2 n)^{1/3} = \frac{m\,k_F}{2\pi^2\hbar^2} \tag{4.148}$$

at the Fermi surface with $k_F = (3\pi^2 n)^{1/3}$ the Fermi wavenumber.

4.6.5 Critical Temperature

The energy gap $\Delta_0(T)$ decreases by increasing the temperature T and it becomes zero at the critical temperature T_c of the superconducting transition, i.e., $\Delta_0(T_c) = 0$. Setting $\beta = \beta_c$ and $\Delta_0 = 0$

in Eq. (4.144) one finds

$$\frac{1}{g_0 \mathscr{D}(0)} = \int_0^{\hbar\omega_D} \frac{\tanh\left(\frac{\beta_c}{2}\xi\right)}{\xi} \, d\xi = \int_0^{\hbar\omega_D/(2k_BT_c)} \frac{\tanh(x)}{x} \, dx \, . \tag{4.149}$$

and from it Bardeen, Cooper and Schrieffer in 1957 obtained

$$k_BT_c = 1.13 \; \hbar\omega_D \; \exp\left(-\frac{1}{g_0 \mathscr{D}(0)} \right) , \tag{4.150}$$

that is a very accurate approximate result for the BCS critical temperature. Remarkably, taking into account Eqs. (4.146) and (4.150), the BCS theory predicts that the zero-temperature energy gap $\Delta_0(0)$ of superconductors in the weak-coupling regime is related to the critical temperature T_c by the simple equation

$$\Delta_0(0) = 1.764 \, k_BT_c \, . \tag{4.151}$$

It is important to stress that the BCS theory describes quite accurately several properties of low-T_c superconductors, mainly metals and alloys. In 1986 Karl Alex Muller and Johannes Georg Bednorz discovered high-T_c superconductors. These ceramic materials (cuprates) can reach the critical temperature of 133 Kelvin. For these high-T_c superconductors, the mechanisms which give rise to pairing of electrons are not yet fully understood.

4.6.6 Condensate Fraction of Cooper Pairs

In the mean-field BCS approximation, we have introduced the uniform and constant energy gap Δ_0. In general, this pairing order parameter is defined as

$$\Delta(\mathbf{r}, \tau) = g_0 \, \langle \psi_\downarrow(\mathbf{r}, \tau) \, \psi_\uparrow(\mathbf{r}, \tau) \rangle \tag{4.152}$$

but at the so-called mean-field level, its spacetime dependence does not appear, i.e., $\Delta(\mathbf{r}, \tau) = \Delta_0$. We have seen that a nonzero value of this quantity implies the formation of Cooper pairs (two fermions with opposite spins), which give rise to quasi-bosonic particles of spin zero. As suggested by Chen Ning Yang in 1962, the condensate wavefunction of these Cooper pairs is given by

$$\Xi(\mathbf{r}, \mathbf{r}', \tau) = \langle \psi_\downarrow(\mathbf{r}, \tau) \, \psi_\uparrow(\mathbf{r}', \tau) \rangle \, . \tag{4.153}$$

This two-particle wave function is strictly related to the largest eigenvalue N_0 of the two-body density matrix of the system. N_0 gives the number of Fermi pairs (Cooper pairs) in the lowest state, i.e., the condensate number of Fermi pairs, and it can be written as

$$N_0 = \int |\Xi(\mathbf{r}, \mathbf{r}', \tau)|^2 \, d^3\mathbf{r} \, d^3\mathbf{r}' \, . \tag{4.154}$$

A finite value for the condensate fraction $f = N_0/(N/2)$ in the thermodynamic limit $N \to \infty$ implies the so-called off-diagonal-long-range-order.

By using the formalism of functional integration, the condensate wavefunction of Cooper pairs can be written as

$$\Xi(\mathbf{r}, \mathbf{r}', \tau) = \frac{1}{\mathcal{Z}} \int \mathcal{D}[\psi_\sigma, \bar{\psi}_\sigma] \, \psi_\downarrow(\mathbf{r}, \tau) \, \psi_\uparrow(\mathbf{r}', \tau) \, \exp\left\{ -\frac{1}{\hbar} \, S[\psi_\sigma, \bar{\psi}_\sigma] \right\} . \tag{4.155}$$

Inserting the BCS Euclidean action (4.120), after some straightforward but lengthy calculations, one finds that the condensate number of Cooper pairs satisfies this expression

$$N_0 = \sum_{\mathbf{k}} \frac{\Delta_0^2}{4E_k^2} \tanh^2\left(\frac{\beta E_k}{2}\right) . \tag{4.156}$$

For the electrons in a superconducting metal, taking into account Eq. (4.143), one then easily finds

$$n_0 = \frac{1}{2} \mathscr{D}(0) \Delta_0^2 \int_0^{\hbar\omega_D} \frac{\tanh^2\left(\frac{\beta}{2}\sqrt{\xi^2 + \Delta_0^2}\right)}{\xi^2 + \Delta_0^2} \, d\xi \tag{4.157}$$

where $n_0 = N_0/L^3$ is the density of electrons in the condensate. At zero temperature Eq. (4.157) gives

$$n_0(0) = \frac{1}{2} \mathscr{D}(0) \Delta_0(0) \arctan\left(\frac{\hbar\omega_D}{\Delta_0(0)} \right) . \tag{4.158}$$

This expression shows that the condensate density $n(0)$ can be expressed in terms of density of states $\mathscr{D}(0)$, energy gap $\Delta_0(0)$ and Debye energy $\hbar\omega_D$. Consequently, the zero-temperature condensate fraction $f(0) = n_0(0)/(n/2)$ is given by

$$f(0) = \frac{1}{2} \frac{\mathscr{D}(0)}{n} \Delta_0(0) \arctan\left(\frac{\hbar\omega_D}{\Delta_0(0)} \right) , \tag{4.159}$$

where n is the density of conduction electrons. Moreover, under the condition $\Delta_0(0) \ll \hbar\omega_D$, the condensate density can be written as

$$n_0(0) = \frac{\pi}{4}\mathscr{D}(0)\Delta_0(0) \, . \tag{4.160}$$

We stress that in many real superconductors the simple mean-field BCS approximation is not accurate, and one has to take into account the retarded electron-electron interaction via phonons and also the Coulomb repulsion. The results obtained above by using the mean-field BCS theory are reliable in the weak-coupling regime, i.e., for $\Delta_0(0) \ll \hbar\omega_D$, where $g\mathscr{D}(0) \leq 0.3$. Therefore, we will continue our analysis of the condensate fraction only for a class of superconductors which satisfy this condition.

Within the free-electron Sommerfeld approximation, the energy spectrum ϵ_k of conduction electrons has a simple quadratic behavior $\epsilon_k = \hbar^2k^2/(2m^*)$, but m^*, the effective mass of electrons, does not coincide with the physical mass m. The free particle density of states $N_{free}(0)$ is related to the total density of conduction electrons by $n = 4N_{free}(0)\mu/3$, and the zero-temperature condensate fraction reads $f(0) = 3\pi\Delta_0(0)/(8\mu)$. To get a better estimate, we correct the free electron value of $\mathscr{D}(0)$ by an effective mass as obtained from specific heat measurements, i.e., we use the expression $\mathscr{D}(0) = (m^*/m)N_{free}(0)$.

In the first two columns of Table 4.1, we show the zero-temperature condensate density n_0 and condensate fraction $f(0)$ of simple metals obtained from Eqs. (4.158) and (4.159) by using experimental data of $\Delta_0(0)$ and ω_D. In the third column we report the electron-phonon strength $g_0\mathscr{D}(0)$ calculated with Eq. (4.145) knowing $f(0)$. The table shows that indeed these simple metals are all in the weak-coupling regime. For completeness, in the fourth column we insert the theoretical determination (see Eq. (4.150)) of the critical temperature T_c, which is very reliable for these simple metals, when compared with the experimental data. Let us now consider the behavior of the condensate density n_0 at finite temperature T. Under the condition $\hbar\omega_D \gg k_BT_c$, from Eq. (4.144) one finds that near T_c the energy gap goes to zero according to the

Table 4.1
BCS predictions for weak-coupling superconducting metals: n_0 is the zero-temperature condensate density, $f(0) = n_0(0)/(n/2)$ is the zero-temperature condensate fraction, $g_0\mathscr{D}(0)$ is the electron-phonon strength, and T_c is the critical temperature. Experimental data taken from N.W. Ashcroft and N.D. Mermin, Solid State Physics (Holt-Sounders, Philadelphia, 1976) and J.P. Carbotte, Rev. Mod. Phys. **62**, 1027 (1990).

	$n_0(0)$ [10^{-33} m^{-3}]	$f(0)$ [10^{-5}]	$g_0\mathscr{D}(0)$	T_c [K]
Cd	4.18	0.9	0.179	0.51
Zn	7.72	1.2	0.172	0.79
Al	22.4	2.5	0.168	1.15
Tl	31.0	5.9	0.263	2.43
In	62.1	10.8	0.267	3.46
Sn	62.5	8.4	0.254	3.68

power law

$$\Delta_0(T) = 3.06\ k_B T_c \left(1 - \frac{T}{T_c}\right)^{1/2} . \tag{4.161}$$

Instead, for the condensate density $n_0(T)$, from Eq. (4.157) and the previous expression, we find near T_c

$$n_0(T) = 0.43\,\mathscr{D}(0)\frac{\Delta_0(T)^2}{k_B T_c} = 4.03\,\mathscr{D}(0) k_B T_c \left(1 - \frac{T}{T_c}\right) . \tag{4.162}$$

Taking into account Eq. (4.159), the BCS theory predicts that the zero-temperature condensate density in superconductors in the weak-coupling regime can be written as

$$n_0(0) = 1.39\,\mathscr{D}(0) k_B T_c . \tag{4.163}$$

4.6.7 Hubbard-Stratonovich Transformation

The Hubbard–Stratonovich transformation is an exact mathematical procedure developed by Ruslan Stratonovich and John

Hubbard in the 1950s. In the context of superconductivity, the Lagrangian density, which is quartic in the fermionic fields, can be rewritten as a quadratic form by introducing an auxiliary field. This procedure is known as the Hubbard–Stratonovich transformation. In particular, given the fermionic partition function

$$\mathcal{Z} = \int \mathcal{D}[\bar{\psi}_\sigma, \psi_\sigma] \, \exp\left\{ -\frac{1}{\hbar} \, S_E[\psi_\sigma, \bar{\psi}_\sigma] \right\} , \tag{4.164}$$

with

$$\begin{aligned} S_E[\bar{\psi}_\sigma, \psi_\sigma] \; &= \; \int_0^{\hbar\beta} d\tau \int d^3\mathbf{r} \, \left\{ \sum_{\sigma=\uparrow,\downarrow} \bar{\psi}_\sigma \left[\hbar\partial_\tau - \frac{\hbar^2}{2m}\nabla^2 - \mu \right] \psi_\sigma \right. \\ & \left. \quad - g_0 \, \bar{\psi}_\uparrow \, \bar{\psi}_\downarrow \, \psi_\downarrow \, \psi_\uparrow \right\} \end{aligned} \tag{4.165}$$

and $g_0 > 0$, it can be rewritten as

$$\mathcal{Z} = \mathcal{N} \int \mathcal{D}[\bar{\psi}_\sigma, \psi_\sigma] \mathcal{D}[\Delta, \Delta^*] \, \exp\left\{ -\frac{1}{\hbar} \, S_e[\bar{\psi}_\sigma, \psi_\sigma, \Delta, \Delta^*] \right\} , \tag{4.166}$$

where

$$\begin{aligned} S_e[\bar{\psi}_\sigma, \psi_\sigma, \Delta, \Delta^*] = \int_0^{\hbar\beta} d\tau \int d^3\mathbf{r} \, \Big\{ \sum_{\sigma=\uparrow,\downarrow} \bar{\psi}_\sigma \left[\hbar\partial_\tau - \frac{\hbar^2}{2m}\nabla^2 - \mu \right] \psi_\sigma \\ - \Delta^* \, \psi_\downarrow \, \psi_\uparrow - \Delta \bar{\psi}_\uparrow \, \bar{\psi}_\downarrow + \frac{|\Delta|^2}{g_0} \Big\} , \end{aligned} \tag{4.167}$$

with $\Delta(\mathbf{r}, \tau)$ the auxiliary complex scalar field and $\mathcal{N}$ a normalization constant that disappears when one calculates thermal (and quantum) averages. Please remember that in the BCS Lagrangian of Eq. (4.108), the complex order parameter Δ_0 is constant and uniform. Instead, here $\Delta(\mathbf{r}, \tau)$ is a complex field of both space and time. In other words, the BCS Lagrangian density (4.108) is exact if the space-time dependent pairing $\Delta(\mathbf{r}, \tau)$ is treated as an independent bosonic complex field with a functional integration over it.

One realizes that the Hubbard-Stratonovich transformation is based on this equation

$$e^{-\frac{1}{\hbar}\int_0^{\hbar\beta} d\tau \int d^3\mathbf{r}\, \left(-g_0\,\bar{\psi}_\uparrow\,\bar{\psi}_\downarrow\,\psi_\downarrow\,\psi_\uparrow\right)}$$
$$= \mathcal{N}\int \mathcal{D}[\Delta,\Delta^*]\; e^{-\frac{1}{\hbar}\int_0^{\hbar\beta} d\tau \int d^3\mathbf{r}\left(-\Delta^*\psi_\downarrow\psi_\uparrow - \Delta\bar{\psi}_\uparrow\bar{\psi}_\downarrow + \frac{|\Delta|^2}{g_0}\right)} . \quad (4.168)$$

We observe that

$$e^{g_0\bar{A}A} = \mathcal{N}\int \frac{dz^* dz}{2\pi i} e^{(z^*A + z\bar{A} - \frac{|z|^2}{g_0})} , \quad (4.169)$$

that is nothing else than Eq. (4.6). Let us prove this formula again in a slightly different way. We notice that

$$g_0\left|\frac{z}{g_0} - A\right|^2 = \frac{|z|^2}{g_0} - z^*A - z\bar{A} + g_0\bar{A}A \quad (4.170)$$

and consequently

$$\begin{aligned}\int \frac{dz^* dz}{2\pi i} e^{(z^*A + z\bar{A} - \frac{|z|^2}{g_0})} &= e^{g_0\bar{A}A}\int \frac{dz dz^*}{2\pi i} e^{-g_0|\frac{z}{g_0} - A|^2} \\ &= e^{g_0\bar{A}A} g_0^2 \int \frac{d\zeta d\zeta^*}{2\pi i} e^{-g_0|\zeta|^2} \\ &= e^{g_0\bar{A}A}\, g_0 \, , \end{aligned} \quad (4.171)$$

setting $\zeta = z/g_0 - A$. It follows that, in this case, the normalization constant is $\mathcal{N} = 1/g_0$.

More generally, considering a complex vectors $\mathbf{z} = (z_1, ..., z_N)$ and $\mathbf{A} = (A_1, ..., A_N)$, it is not difficult to prove that

$$e^{g_0\bar{\mathbf{A}}\cdot\mathbf{A}} = \mathcal{N}\int \frac{d^N\mathbf{z}^* \cdot d^N\mathbf{z}}{2\pi i} e^{(\mathbf{z}^*\cdot\mathbf{A} + \mathbf{z}\cdot\bar{\mathbf{A}} - \frac{|\mathbf{z}|^2}{g_0})} \quad (4.172)$$

with $\mathcal{N} = 1/g_0^N$. See also Eq. (4.10). Finally, it is straightforward to extend this approach to complex fields obtaining the required result.

It is very important to stress that the Euclidean action (4.167) can be used to study beyond-mean-field corrections encoded in the pairing field $\Delta(\mathbf{r},\tau)$. For instance, writing

$$\Delta(\mathbf{r},\tau) = \Delta_0 + \eta(\mathbf{r},t) \quad (4.173)$$

and treating $\eta(\mathbf{r}, \tau)$ as a small fluctuation around the saddle-point configuration Δ_0, one can derive and investigate the Gaussian terms of $\eta(\mathbf{r}, \tau)$ which improve the BCS grand potential.

4.7 BCS-BEC Crossover

In 1969 David Eagles suggested the possibility of having for a fermionic superconductor the crossover from weakly-bound BCS-like Cooper pairs to the Bose-Einstein condensation (BEC) of strongly-bound two-electron states. In 1980 Anthony Leggett showed that, in the context of a dilute ultracold Fermi gas, the scattering theory plays an essential role in the description of this BCS-BEC crossover. Quite remarkably, these theoretical predictions of Eagles and Leggett have been experimentally confirmed in the last twenty years with ultracold atoms and, very recently, also with some superconducting material.

Let us consider again the gap equation in three spatial dimensions

$$\frac{1}{g_0} = \frac{1}{L^3} \sum_{|\mathbf{k}|<\Lambda} \frac{\tanh\left(\frac{\beta E_k}{2}\right)}{2E_k} \tag{4.174}$$

and the number equation

$$n = \frac{1}{L^3} \sum_{|\mathbf{k}|<\Lambda} \left(1 - \frac{\frac{\hbar^2 k^2}{2m} - \mu}{E_k}\right) \tanh\left(\frac{\beta E_k}{2}\right) \tag{4.175}$$

of the mean-field BCS theory, where the ultraviolet cutoff Λ has been inserted to avoid the divergence in the gap equation. Leggett suggested a different regularization procedure, relating the physical s-wave scattering length a_F between fermionic atoms to the actual interatomic potential to the strength g_0 of the contact interaction:

$$\frac{1}{g_0} = -\frac{m}{4\pi\hbar^2 a_F} + \frac{1}{L^3} \sum_{|\mathbf{k}|<\Lambda} \frac{m}{\hbar^2 k^2} . \tag{4.176}$$

This equation, found by Nikolaj Bogoliubov in 1947, can be deduced from the Lippman-Schwinger equation for the T-matrix of scattering theory discussed in the previous chapter. Indeed, Eq. (4.176) is nothing else than Eq. (3.155) with the inclusion of the ultraviolet cutoff Λ. Inserting Eq. (4.176) into Eq. (4.174), we get

$$-\frac{m}{4\pi\hbar^2 a_F} = \frac{1}{L^3}\sum_{|\mathbf{k}|<\Lambda}\left(\frac{\tanh\left(\frac{\beta E_k}{2}\right)}{2E_k} - \frac{m}{\hbar^2 k^2}\right) . \tag{4.177}$$

Quite remarkably, in the limit $\Lambda \to +\infty$, this equation is not divergent because the two terms on the right-hand side compensate each other for large momenta. Thus, we can safely write in the continuum

$$-\frac{m}{4\pi\hbar^2 a_F} = \frac{1}{2\pi^2}\int_0^{+\infty} dk\, k^2 \left(\frac{\tanh\left(\frac{\beta E_k}{2}\right)}{2E_k} - \frac{m}{\hbar^2 k^2}\right) . \tag{4.178}$$

This is the regularized gap equation for the BCS-BEC crossover. The key result is that this equation works both for negative and positive values of the scattering length a_F. The BCS regime corresponds to a small and negative value of a_F while the BEC regime is associated to a small and positive a_F. In between, when a_F is extremely large, there is the unitarity limit where, strictly speaking, one has $a_F = \pm\infty$. Specifically, after introducing the adimensional interaction strength

$$y = \frac{1}{k_F a_F} \tag{4.179}$$

with $k_F = (3\pi^2 n)^{1/3}$ the Fermi wavenumber, the three relevant limiting cases are: (i) the BCS regime, where $y \lll -1$; (ii) the unitarity limit, with $y = 0$ (thus $a_F = \pm\infty$); (iii) the BEC regime, where $y \gg 1$, $\mu/\Delta_0 \ll -1$. In Fig. 4.1 we plot the phase diagram of the BCS-BEC crossover, i.e., thermal energy $k_B T$ in units of the Fermi energy ϵ_F as a function of the inverse interaction strength $1/(k_F a_F)$. The mean-field critical temperature T^* (dashed line) is reliable only in the BCS regime of the crossover. To get the solid line one must also include the effect of pairing fluctuations.

At zero temperature, by solving the renormalized gap equation coupled to the number equation, one finds simple analytical mean-field results for the chemical potential μ, the energy gap Δ_0, and

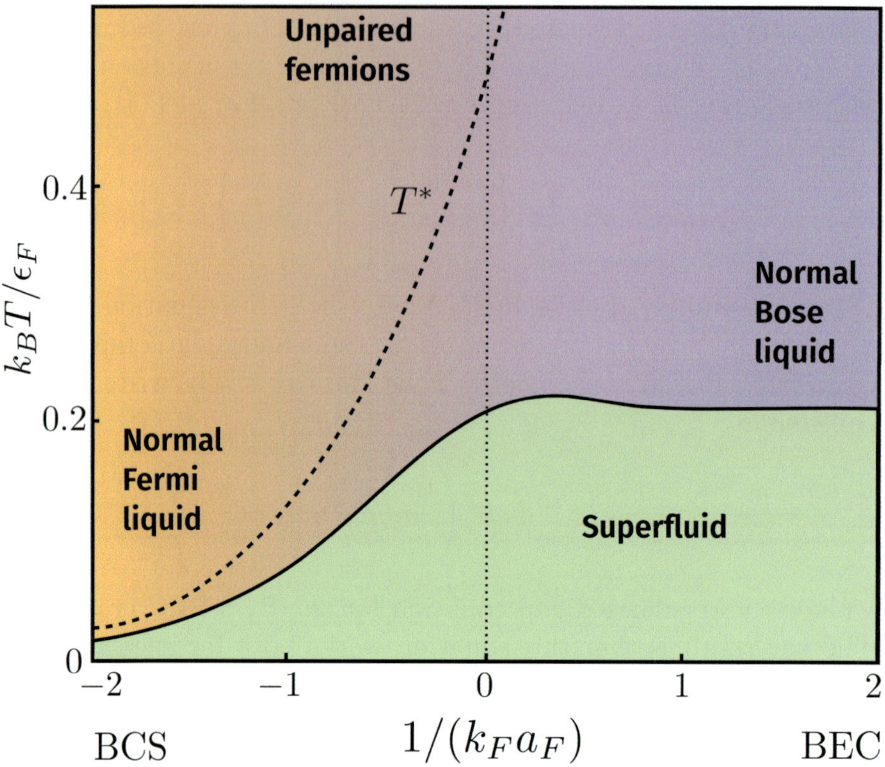

Figure 4.1
Phase diagram of the BCS-BEC crossover. T^* is the mean-field critical temperature of the superfluid-to-normal phase transition. a_F is the fermion-fermion scattering length.

the condensate fraction n_0/n in these three limiting cases. The main results are the following:
(i) *BCS regime*: μ approaches the non-interacting Fermi energy $\epsilon_F = \hbar^2 k_F^2/(2m)$. One finds that the condensate density is given by

$$n_0 \simeq \frac{m k_F}{8\pi\hbar^2}\,\Delta_0 = \frac{3\pi}{2e^2}\,n\,\exp\!\left(\frac{\pi}{2}\,y\right)\,, \tag{4.180}$$

with an exponentially small energy gap $\Delta_0 = 8e^{-2}\,\epsilon_F\,\exp\{(y\,\pi/2)\}$. Here the size of weakly bound Cooper pairs exceeds the typical interparticle spacing $k_{\rm F}^{-1}$.

(ii) *Unitarity limit*: one finds $\mu \simeq 0.59\epsilon_F$ and $\Delta_0 \simeq 0.69\epsilon_F$, and the condensate fraction is

$$\frac{2\,N_0}{N} \simeq 0.6994\,. \tag{4.181}$$

Here the size of Cooper pairs is of the order of the interparticle spacing $k_{\rm F}^{-1}$.
(iii) *BEC regime*: the condensate fraction approaches the ideal Bose gas value

$$\frac{2\,N_0}{N} \simeq 1\,, \tag{4.182}$$

with all pairs (molecular dimers) moving into the condensate. Here the size of Cooper pairs is much smaller than the interparticle spacing $k_{\rm F}^{-1}$. The experimental achievement of the BCS-BEC crossover with ultracold fermionic atoms was obtained in 2003 by using the technique of Feschbach resonances, where the scattering length a_F of the atoms in a confined ultracold fermionic gas is tuned at will by an external constant magnetic field $\mathbf{B}$. In particular, near a Feshbach resonance, where a_F diverges, the behavior of a_F as a function of the magnetic field B is well captured by

$$a_F = a_{bg}\left[1 + \alpha_1(B - B_0)\right]\left[1 + \frac{\alpha_2}{B - B_0}\right]\,, \tag{4.183}$$

where B_0 is the critical magnetic field at which the resonance appears, a_{bg} is the background scattering length, and α_1 and α_2 are two phenomenological parameters.

4.7.1 Two-Dimensional Crossover

The BCS-BEC crossover was observed not only with fermionic ^{6}Li atoms in a three-dimensional configuration (2003) but also in a quasi two-dimensional one (2014). Quite generally, in two spatial dimensions, a bound-state energy ϵ_B exists for any value of the interaction strength g. For the contact potential, the bound-state equation is given by

$$-\frac{1}{g} = \frac{1}{L^2}\sum_{\mathbf{k}} \frac{1}{\frac{\hbar^2 k^2}{2m} + \epsilon_B}\,, \tag{4.184}$$

and then subtracting this equation from the gap equation, one obtains a regularized gap equation

$$\sum_{\mathbf{k}} \left(\frac{1}{\frac{\hbar^2 k^2}{2m} + \epsilon_B} - \frac{1}{2E_k} \right) = 0 \, . \tag{4.185}$$

In the two-dimensional continuum limit $\sum_{\mathbf{k}} \to L^2/(2\pi)^2 \int d^2\mathbf{k} \to V/(2\pi) \int k dk$, taking into account the functional dependence of the amplitudes u_k and v_k on μ and Δ_0, Eq. (4.185) gives

$$\epsilon_B = \Delta_0 \left(\sqrt{1 + \frac{\mu^2}{\Delta_0^2}} - \frac{\mu}{\Delta_0} \right) , \tag{4.186}$$

while the number equation becomes

$$n = \frac{N}{L^2} = (\frac{m}{2\pi\hbar^2}) \Delta_0 \left(\frac{\mu}{\Delta_0} + \sqrt{1 + \frac{\mu^2}{\Delta_0^2}} \right) . \tag{4.187}$$

Combining Eqs. (4.186) and (4.187), one finds the quite elegant mean-field result

$$\mu = \epsilon_F - \frac{1}{2}\epsilon_B \tag{4.188}$$

with $\epsilon_F = \pi\hbar^2 n/m$ the 2D Fermi energy of non-interacting fermions. Also for the condensate fraction, one obtains an analytical expression in the two-dimensional continuum limit, namely

$$n_0 = \frac{N_0}{L^2} = \frac{1}{4} \int \frac{d^2k}{2\pi} \frac{\Delta_0^2}{\left(\frac{\hbar^2k^2}{2m} - \mu\right)^2 + \Delta_0^2} , \tag{4.189}$$

from which

$$n_0 = \frac{1}{4} (\frac{m}{2\pi\hbar^2}) \Delta_0 \left(\frac{\pi}{2} + \arctan\left(\frac{\mu}{\Delta_0}\right) \right) . \tag{4.190}$$

Finally, by using Eq. (4.187) and Eq. (4.190), one gets a simple formula for the condensed fraction

$$\frac{n_0}{n} = \frac{1}{4} \frac{\frac{\pi}{2} + \arctan\left(\frac{\mu}{\Delta_0}\right)}{\frac{\mu}{\Delta_0} + \sqrt{1 + \frac{\mu^2}{\Delta_0^2}}} , \tag{4.191}$$

In Eq. (4.191) the condensate fraction depends only on the parameter $x_0 = \mu/\Delta_0$. In the weakly-bound BCS regime ($x_0 \gg 1$), the condensed fraction n_0/n goes to zero, while in the strongly-bound BEC regime ($x_0 \ll -1$), the condensed fraction n_0/n goes to 1/2, i.e., all the $N/2$ Fermi pairs belong to the Bose-Einstein condensate (Salasnich, 2007).

At finite temperature, due to the Mermin-Wagner theorem, there is no Bose-Einstein condensation for a two-dimensional uniform system with short-range interaction. There is, however, the possibility of having superfluidity. In our case, the BCS spectrum

$$E_k = \sqrt{\left(\frac{\hbar^2 k^2}{2m} - \mu\right)^2 + \Delta_0^2} \tag{4.192}$$

of single-particle fermionic excitations is gapped, and this strongly suggests that the two-dimensional attractive Fermi gas is superfluid below a critical temperature. By using the Landau formula, the superfluid density reads

$$n_s^{(0)}(T) = n - \beta \int \frac{d^2q}{(2\pi)^2} \frac{\hbar^2 k^2}{2m} \frac{e^{\beta E_k}}{e^{\beta E_k} + 1} \tag{4.193}$$

where the thermal distribution is that one of fermionic BCS quasiparticles. This is the bare superfluid density which does not include the formation of vortex-antivortex pairs by increasing the temperature and the proliferation of free vortices above the BKT critical temperature. From Eq. (4.193), by applying the Nelson-Kosterlitz condtion

$$k_B T_{BKT} = \frac{\hbar^2 \pi}{8m} n_s^{(0)}(T_{BKT}) \tag{4.194}$$

one gets an implicit formula for the Berezinskii-Kosterlitz-Thouless critical temperature T_{BKT}, i.e.

$$k_B T_{BKT} = n - \frac{1}{2\pi k_B T_{BKT}} \int_0^{+\infty} dk\, q \frac{\hbar^2 k^2}{2m} \frac{e^{E_k/(k_B T_{BKT})}}{e^{E_k/(k_B T_{BKT})} + 1}. \tag{4.195}$$

In general, the Nelson-Kosterlitz criterion overestimates the BKT temperature with respect to the renormalization group equations, this effect being particularly relevant in the intermediate regime of the crossover. Moreover, in the BEC side of the crossover, it is relevant to include also the collective modes (density oscillations), which can be extracted from the BCS action functional by including Gaussian fluctuations (Bighin and Salasnich, 2016).

It is very important to underline that the BCS-BEC crossover can be achieved also with two-dimensional superconducting materials by tuning the carrier density and the geometry of the samples. For instance, in 2021 Yoshihiro Iwasa and collaborators reported a comprehensive characterization of the BKT superconducting critical temperature across the BCS-BEC crossover.

4.8 Ginzburg-Landau Theory

A breakthrough in the understanding of phase transitions occurred in 1937, the year in which Lev Landau observed that in a phase transition, a more symmetrical phase can be distinguished compared to the other less symmetrical one. Typically, the phase with the highest symmetry is on the hotter side of the transition and is called *disordered phase*; the one with the lowest symmetry, placed on the colder side, is called *ordered phase*. For example, in the liquid-solid transition of water, the liquid phase is composed of molecules arranged in a disorderly but homogeneous manner. Liquid has continuous translational symmetry: every point within the fluid has the same properties as any other point. Instead, it is formed in the solid phase as a well-ordered crystalline structure, where the atoms are arranged in a regular grid. The arrangement of the atoms of the solid appears different if observed from different points, unless these points are not separated by an integer number of lattice steps. Another example is the ferromagnetic transition. In this case the symmetry that is broken is the rotational invariance of the magnetic moments of atoms. In the ferromagnetic phase, the symmetry breaks due to the formation of magnetic domains containing atoms with magnetic moments all aligned. Inside the domain there is therefore a vector of non-zero

Table 4.2
Order parameters for some phase transitions.

Phase transition	Order parameter
ferromagnetic	magnetization
gas-liquid at the critical point	density difference
superfluidity	macroscopic condensate wavefunction
superconductivity	macroscopic wavefunction of Cooper pairs

magnetization which points in a certain direction, chosen spontaneously among the infinite possibilities during the transition: in this case we speak of *spontaneous symmetry breaking.* Notice that a ferromagnetic material has many magnetic domains, each with magnetization $\mathbf{M}_i$ along a different direction. To have a total magnetization $\mathbf{M} = \sum_i \mathbf{M}_i$ other than zero, it is necessary to apply an external magnetic field $\mathbf{B}$ which orients all magnetic vectors $\mathbf{M}_i$ along the direction of $\mathbf{B}$.

According to Landau, when the symmetry is broken and the system goes from the disordered phase to the ordered one, you need to introduce another variable to describe the thermodynamics of the ordered phase. This additional variable is called *order parameter* and is normally a variable in thermodynamics that can be measured experimentally. In the ferromagnetic phase the order parameter is the magnetization of a domain, whose value is different from zero when the system is cooled below the Curie point. In the gas-liquid transition in proximity of the critical point, the order parameter It is the difference in density between the gaseous state and the liquid one. In the superfluid phase, the order parameter is the Bose-Einstein condensate wave function, where a macroscopic number of bosonic particles occupies the same single particle state. In superconductivity the order parameter is the wave function of Cooper pairs: due to the interaction with the ionic lattice, electrons with opposite spin form bound states (Cooper pairs) whose presence gives rise to superconductivity.

In 1950, seven years before the BCS theory, Lev Landau and Vitaly Ginzburg proposed a phenomenological approach to describe the superconducting phase transition based on the complex order parameter $\Psi(\mathbf{r})$. In 1959 Lev Gor'kov showed that from the mean-field BCS theory, one deduces the Ginzburg-Landau theory and that the energy gap Δ_0 can be identified as the uniform order parameter Ψ_0. We notice, however, that the Ginzburg-Landau theory is somehow better than the mean-field BCS theory because, in general, the Ginzburg-Landau order parameter $\Psi(\mathbf{r})$ is not uniform, while the mean-field BEC energy gap Δ_0 is assumed to be uniform. From this point of view, the Ginzburg-Landau order parameter $\Psi(\mathbf{r})$ plays a similar role to the pairing field $\Delta(\mathbf{r},\tau)$, that is, the Hubbard-Stratonovich auxiliary bosonic field, introduced in the previous subsection. In particular, $\Psi(\mathbf{r})$ can be identified as the high-temperature pairing field $\Delta(\mathbf{r})$ where the dependence on the imaginary time τ is negligible.

The main idea of Ginzburg and Landau was that, close to the critical temperature, the Helmholtz free energy of a superconducting material can be written as

$$F = F_n + F_s \,, \tag{4.196}$$

where F_n is the contribution due to the normal component and F_s is the contribution due to the emergence of a superconducting complex order parameter $\Psi(\mathbf{r})$ below a critical temperature. Within the Ginzburg-Landau approach, for a D-dimensional system of area L^D, the super component F_s is given by

$$F_s = \int_{L^D} d^D\mathbf{r} \left\{ a(T)\,|\Psi(\mathbf{r})|^2 + \frac{b}{2}\,|\Psi(\mathbf{r})|^4 + \frac{\hbar^2}{2m^*}\,|\nabla\Psi(\mathbf{r})|^2 \right\} \,, \tag{4.197}$$

where

$$a(T) = \alpha\, k_B\,(T - T_c) \tag{4.198}$$

is a parameter which depends on the temperature T (k_B is the Boltzmann constant) and becomes zero at the mean-field critical temperature T_c, while $b > 0$ and $m^* > 0$ are temperature-independent phenomenological parameters.

The partition function Z of the system is given by

$$\mathcal{Z} = e^{-\beta F_n} \, \mathcal{Z}_s \tag{4.199}$$

where

$$\mathcal{Z}_s = \int d[\Psi^*, \Psi] \, e^{-\beta F_s[\Psi^*, \Psi]} \; , \tag{4.200}$$

with $\beta = 1/(k_B T)$, and the thermal average of an observable A that is a functional of $\Psi(\mathbf{r})$ and $\Psi^*(\mathbf{r})$ reads

$$\langle A \rangle = \frac{1}{\mathcal{Z}_s} \int d[\Psi^*, \Psi] \, A[\Psi^*, \Psi] \, e^{-\beta F_s[\Psi, \Psi^*]} \; . \tag{4.201}$$

Notice that in these functional integrals, the complex field $\Psi(\mathbf{r})$ does not depend on time. As shown in the previous chapter, they are fully reliable at high temperatures where quantum fluctuations are negligible. Assuming a real and uniform order parameter, i.e.

$$\Psi(\mathbf{r}) = \Psi_0 \; , \tag{4.202}$$

the energy functional (4.197) with Eqs. (4.198) and (4.202) becomes

$$\frac{F_{s0}}{L^D} = a(T) \, \Psi_0^2 + \frac{b}{2} \, \Psi_0^4 \; . \tag{4.203}$$

Minimizing F_{s0} with respect to Ψ_0, one immediately finds

$$a(T) \, \Psi_0 + b \, \Psi_0^3 = 0 \; , \tag{4.204}$$

and consequently

$$\Psi_0 = \begin{cases} 0 & \text{for } T \geq T_c \\ \sqrt{-\frac{a(T)}{b}} & \text{for } T < T_c \end{cases} \; . \tag{4.205}$$

Thus, the uniform order parameter Ψ_0 becomes different from zero only below the mean-field critical temperature T_c, where, by definition, T_c is such that

$$a(T_c) = 0 \; . \tag{4.206}$$

For $0 \leq T < T_c$ the free energy F_{s0} is then given by

$$\frac{F_{s0}}{L^D} = -\frac{a(T)^2}{b} = -\frac{\alpha^2 k_B^2}{b} (T_c - T)^2 \; , \tag{4.207}$$

while for $T \geq T_c$ we have $F_{s0} = 0$.

As previously said, Gor'kov showed that the phenomenological Ginzburg-Landau theory can be deduced from the microscopic BCS theory. In particular, the order parameter Ψ_0 can be identified with the BCS energy gap Δ_0, while the coefficients α, b, and m^* of the Ginzburg-Landau energy functional (4.197) are directly related to the parameters of the BCS Lagrangian density (4.108), taking into account Eqs. (4.146), (4.150), (4.198), and (4.205).

4.8.1 Superconductors in a Magnetic Field

In general, to describe superconductors, we must also take into account the coupling with the electromagnetic field. The minimal coupling reads

$$-i\hbar\boldsymbol{\nabla} \to -i\hbar\boldsymbol{\nabla} - q\mathbf{A}(\mathbf{r}) \tag{4.208}$$

where $\mathbf{A}(\mathbf{r})$ is the vector potential, q is the charge of the composite bosonic-like particles described by the field $\Psi(\mathbf{r})$, and $i = \sqrt{-1}$ is the imaginary unit. Thus, the Ginzburg-Landau functional becomes

$$\begin{aligned} F_s = \int_{L^D} d^D\mathbf{r}\Big\{ & a(T)\,|\Psi(\mathbf{r})|^2 + \frac{b}{2}\,|\Psi(\mathbf{r})|^4 \\ & + \frac{\hbar^2}{2m^*}\,\Big|\left(\boldsymbol{\nabla} - i\frac{q}{\hbar}\mathbf{A}(\mathbf{r})\right)\Psi(\mathbf{r})\Big|^2 + \frac{1}{2\mu_0}|\mathbf{B}(\mathbf{r})|^2\Big\}\,, \end{aligned} \tag{4.209}$$

where

$$\mathbf{B}(\mathbf{r}) = \boldsymbol{\nabla} \wedge \mathbf{A}(\mathbf{r}) \tag{4.210}$$

is the space-dependent magnetic field. The last term in Eq. (4.209) takes into account the free magnetic energy of the system with μ_0 the magnetic permeability in the vacuum. Notice that, in principle, an electric component $\int d^3\mathbf{r}\frac{1}{2}\epsilon_0|\mathbf{E}(\mathbf{r})|^2$ should be added to the Ginzburg-Landau free energy, where ϵ_0 is the dielectric parameter in the vacuum and

$$\mathbf{E}(\mathbf{r}) = -\boldsymbol{\nabla}\phi(\mathbf{r}) - \frac{\partial}{\partial t}\mathbf{A}(\mathbf{r}) = -\boldsymbol{\nabla}\phi(\mathbf{r}) \tag{4.211}$$

is the electric field, with $\phi(\mathbf{r})$ the electromagnetic scalar potential. Here $\partial\mathbf{A}(\mathbf{r})/\partial t = 0$ because in our specific problem, the vector potential $\mathbf{A}(\mathbf{r})$ is time independent. The minimization of the functional $F_s = F_s[\Psi(\mathbf{r}), \mathbf{A}(\mathbf{r})]$ with respect to $\psi^*(\mathbf{r})$ gives the so-called Ginzburg-Landau equation

$$\left[-\frac{\hbar^2}{2m^*} \left(\boldsymbol{\nabla} - i\frac{q}{\hbar}\mathbf{A}(\mathbf{r}) \right)^2 + a(T) + b\,|\Psi(\mathbf{r})|^2 \right] \Psi(\mathbf{r}) = 0\,, \quad (4.212)$$

that is a nonlinear Schrödinger equation with cubic nonlinearity for the order parameter $\Psi(\mathbf{r})$, which contains the minimal coupling with the vector potential $\mathbf{A}(\mathbf{r})$.

4.8.2 Supercurrent

The supercurrent charge density $\mathbf{j}_s(\mathbf{r})$ can be obtained by considering the minimization of the functional $F_s = F_s[\Psi(\mathbf{r}), \mathbf{A}(\mathbf{r})]$ with respect to $\mathbf{A}(\mathbf{r})$. In this way one finds

$$\begin{aligned} \frac{1}{\mu_0}\boldsymbol{\nabla} \wedge \mathbf{B}(\mathbf{r}) &= i\frac{q\hbar^2}{2m^*} \left(\Psi^*(\mathbf{r})\boldsymbol{\nabla}\Psi(\mathbf{r}) - \Psi(\mathbf{r})\boldsymbol{\nabla}\Psi^*(\mathbf{r}) \right) \\ &\quad - \frac{k^2}{m^*}|\Psi(\mathbf{r})|^2\mathbf{A}(\mathbf{r}). \end{aligned} \quad (4.213)$$

Remarkably, Eq. (4.213) can be rewritten as the familiar Ampere equation

$$\boldsymbol{\nabla} \wedge \mathbf{B}(\mathbf{r}) = \mu_0\,\mathbf{j}_s(\mathbf{r}) \quad (4.214)$$

setting

$$\mathbf{j}_s(\mathbf{r}) = i\frac{q\hbar}{2m^*} \left(\Psi^*(\mathbf{r})\boldsymbol{\nabla}\Psi(\mathbf{r}) - \Psi(\mathbf{r})\boldsymbol{\nabla}\Psi^*(\mathbf{r}) \right) - \frac{k^2}{m^*}|\Psi(\mathbf{r})|^2\mathbf{A}(\mathbf{r})\,, \quad (4.215)$$

which is identified as the supercurrent charge density.

Taking into account that the order parameter $\Psi(\mathbf{r})$ describes bosons with effective mass m^* and effective charge q, which are made of Cooper pairs (two electrons with opposite spin), we set

$$q = -2e \quad (4.216)$$

$$m^* = 2m_e \quad (4.217)$$

where $-e$ is the negative electric charge of the electron (with $e >$ the electric charge of the proton) and m_e the mass of the electron.

We also introduce, in full generality, the superfluid local number density of electrons as

$$n_s(\mathbf{r}) = 2|\Psi(\mathbf{r})|^2 . \tag{4.218}$$

Assuming a real and uniform order parameter, see Eq. (4.202), the supercurrent is strongly simplified and reads

$$\mathbf{j}_s(\mathbf{r}) = -\frac{k^2}{m^*}\Psi_0^2\mathbf{A}(\mathbf{r}) . \tag{4.219}$$

Clearly, from Eq. (4.218), in the case of a uniform and real order parameter, we have

$$\Psi_0(T)^2 = \frac{n_s(T)}{2} . \tag{4.220}$$

In addition, using Eqs. (4.216), (4.217), and (4.219), the supercurrent can be rewritten as

$$\mathbf{j}_s(\mathbf{r}) = -\frac{e^2 n_s(T)}{m_e}\mathbf{A}(\mathbf{r}) . \tag{4.221}$$

This is the London equation, obtained for the first time by the Fritz London and Heinz London in 1935.

4.8.3 London Penetration Depth

The curl of the Ampere equation (4.214) gives

$$-\nabla^2\mathbf{B}(\mathbf{r}) = \mu_0 \, \boldsymbol{\nabla} \wedge \mathbf{j}_s(\mathbf{r}) , \tag{4.222}$$

taking into account that

$$\boldsymbol{\nabla} \wedge (\boldsymbol{\nabla} \wedge \mathbf{B}) = -\nabla^2\mathbf{B} + \boldsymbol{\nabla}(\boldsymbol{\nabla} \cdot \mathbf{B}) = -\nabla^2\mathbf{B} \tag{4.223}$$

due to the magnetic Gauss law

$$\boldsymbol{\nabla} \cdot \mathbf{B} = 0 . \tag{4.224}$$

Inserting Eq. (4.221) into Eq. (4.222), and using Eq. (4.210), we get

$$\nabla^2\mathbf{B}(\mathbf{r}) = \frac{e^2 n_s(T)\mu_0}{m_e}\mathbf{B}(\mathbf{r}) . \tag{4.225}$$

Assuming that $\mathbf{B} = \mathbf{B}(x)$ the previous equation can be written as

$$\frac{\partial^2}{\partial x^2}\mathbf{B}(x) = \frac{e^2 n_s(T)\mu_0}{m_e}\mathbf{B}(x) \tag{4.226}$$

which has the following physically relevant solution for $x \geq 0$:

$$\mathbf{B}(x) = \mathbf{B}(0)\, e^{-x/\lambda_L(T)}\,, \tag{4.227}$$

where

$$\lambda_L(T) = \sqrt{\frac{m_e}{e^2 n_s(T)\mu_0}} \tag{4.228}$$

is the so-called London penetration depth. The meaning of Eq. (4.227) is that inside a superconductor, the magnetic field decays exponentially. This is the Meissner-Ochsenfeld effect: the expulsion of a magnetic field from a superconductor, experimentally observed for the first time in 1933.

4.8.4 Coherence Length and Quantized Vortices

Another very important characteristic length of the Ginzburg-Landau theory is the so-called coherence length

$$\xi(T) = \sqrt{\frac{\hbar^2}{2m^*|a(T)|}}\,. \tag{4.229}$$

This is the distance at which there is a compensation between the gradient term and the linear term of the Ginzburg-Landau equation:

$$|-\frac{\hbar^2}{2m^*}\nabla^2\psi| \simeq \frac{\hbar^2}{2m^*\xi(T)^2}|\psi| = |a(T)||\psi|\,. \tag{4.230}$$

Alexei Abrikosov in 1957 showed that the coherence length $\xi(T)$ is nothing else than the healing length of the quantized vortices of the Ginzburg-Landau equation. Invoking the presence of quantized vortices, Abrikosov explained type-II superconductors, which were discovered by Rjabinin and Shubnikov in 1935.

As shown in Fig. 4.2 the known superconductors can be classified into one of two classes, or types. Superconductors classified

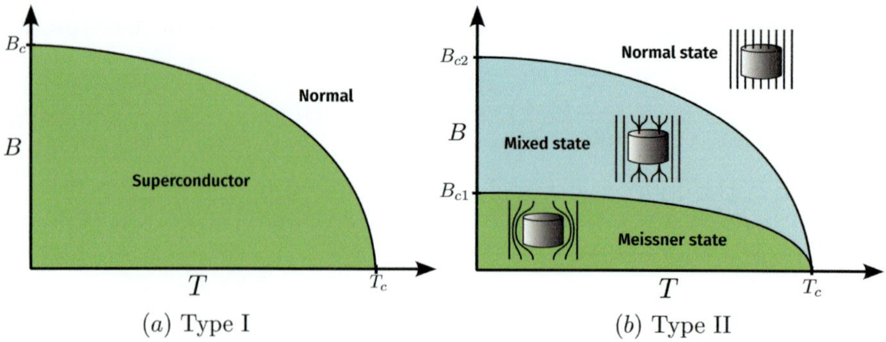

Figure 4.2
Phase diagram of the magnetic field B vs the temperature T for type I and type II superconductors. Figure adapted from R.G. Sharma, Superconductivity. Basics and Applications to Magnets (Springer, 2021).

as type I exhibits a distinct shift from the superconducting state, in which all magnetic flux is released, to the normal state. Type II superconductors behave similarly, eliminating all magnetic field below a certain critical field value and returning to normal at a higher critical field. Nonetheless, the type II superconductor enters a "mixed state" where partial flux penetration occurs when the magnetic field is between these lower and upper critical values.

Following an earlier intuition of Ginzburg and Landau, by solving the Ginzburg–Landau equation, Abrikosov found that a type-II superconductor appears when

$$\frac{\lambda_L(T)}{\xi(T)} > \frac{1}{\sqrt{2}} \simeq 0.7071 \,, \tag{4.231}$$

where $\lambda_L(T)$ is the London penetration depth and $\xi(T)$ is the coherence length. Under this condition, there is the formation of Abrikosov quantized vortices, and the magnetic field can penetrate deep inside the superconductor. Abrikosov found that the vortices arrange themselves into a regular array known as a vortex lattice. A similar analysis was done for the problem of vortex state in a rotating superfluid by Lars Onsager and Richard Feynman. As a final comment, we note that for $T < T_c$ we have

$$\lambda_L(T) = \sqrt{\frac{m_e}{e^2\mu_0 n_s(T)}} = \sqrt{\frac{m_e b}{2e^2\mu_0 \alpha k_B(T_c - T)}} \tag{4.232}$$

and

$$\xi(T) = \sqrt{\frac{\hbar^2}{4m_e|a(t)|}} = \sqrt{\frac{\hbar^2}{4m_e \alpha k_B(T_c - T)}} . \tag{4.233}$$

Then it follows

$$\kappa = \frac{\lambda_L(T)}{\xi(T)} = \frac{m_e}{e\hbar}\sqrt{\frac{2b}{\mu_0}} . \tag{4.234}$$

This means that, within the Ginzburg-Landau theory, the ratio κ is temperature independent. Instead, κ depends on the universal constants $\hbar$, m_e, e, and μ_0 but also on the phenomenological Ginzburg-Landau parameter b.

4.8.5 Thermal Gaussian Fluctuations

We know that the mean-field free energy of Eq. (4.207) is lacking relevant information encoded in the thermal fluctuations of the order parameter $\Psi(\mathbf{r})$. Let us set

$$\Psi(\mathbf{r}) = \Psi_0 + \eta(\mathbf{r}) \tag{4.235}$$

and expand the functional $F[\Psi, \Psi^*]$ around Ψ_0 up to quadratic (Gaussian) order in $\eta(\mathbf{r})$ and $\eta^*(\mathbf{r})$. We find

$$\mathcal{Z}_s = \mathcal{Z}_{s0} \int d[\eta^*, \eta] \; \exp\left\{-\beta F_{sg}[\eta^*, \eta]\right\} , \tag{4.236}$$

where

$$F_{sg}[\eta^*, \eta] = \frac{1}{2\beta} \sum_{\mathbf{k}} (\eta^*_{\mathbf{k}}, \eta_{-\mathbf{k}}) \; \hat{M}_{\mathbf{k}} \begin{pmatrix} \eta_{\mathbf{k}} \\ \eta^*_{-\mathbf{k}} \end{pmatrix} \tag{4.237}$$

is the Gaussian free energy of fluctuations in reciprocal space with $\mathbf{k}$ the D vector denoting the momentum, and

$$\hat{M}_{\mathbf{k}} = \beta \begin{pmatrix} \frac{\hbar^2 k^2}{2m^*} + a(T) + 2b\Psi_0^2 & -b\Psi_0^2 \\ -g\psi_0^2 & \frac{\hbar^2 k^2}{2m^*} + a(T) + 2b\Psi_0^2 \end{pmatrix} \tag{4.238}$$

is the inverse fluctuation propagator. Integrating over the bosonic fields $\eta(\mathbf{k})$ and $\eta^*(\mathbf{k})$ in Eq. (4.236) one finds the Gaussian free energy

$$F_{sg} = \frac{1}{2\beta} \sum_{\mathbf{k}} \ln \mathrm{Det}[\hat{M}_{\mathbf{k}}] = \frac{1}{\beta} \sum_{\mathbf{k}} \ln (\beta E_k) \;, \tag{4.239}$$

where E_k is given by

$$E_k = \sqrt{\left(\frac{\hbar^2 k^2}{2m^*} + a(T) + 2b\Psi_0^2 \right)^2 - b^2 \Psi_0^4} \;. \tag{4.240}$$

Thus, for $T > T_c$ where $\Psi_0 = 0$ we get

$$E_k = \frac{\hbar^2 k^2}{2m^*} + \alpha k_B (T - T_c) \tag{4.241}$$

while for $T < T_c$ where $\Psi_0^2 = \alpha k_B (T_c - T)/b$, we have

$$E_k = \sqrt{\frac{\hbar^2 k^2}{2m^*} \left(\frac{\hbar^2 k^2}{2m^*} + \frac{\alpha k_B}{b} (T_c - T) \right)} \;, \tag{4.242}$$

that is a sort of Bogoliubov-like spectrum. In the continuum limit the free energy becomes

$$\frac{F_s}{L^D} = \frac{F_{s0}}{L^D} + k_B T \int_{|\mathbf{k}|<\Lambda} \frac{d^D\mathbf{k}}{(2\pi)^D} \ln \left(\frac{E_k}{k_B T} \right) \;, \tag{4.243}$$

where

$$\Lambda = \frac{1}{\xi(0)} = \sqrt{\frac{\hbar^2}{2m^* \alpha k_B T_c}} \tag{4.244}$$

is an ultraviolet cutoff related to the lowest value of the healing length $\xi(T)$, that is the healing length $\xi(0)$ at zero temperature. The superconducting free energy F_s can be calculated numerically from Eq. (4.243). Having it one can then easily determine the superconducting entropy

$$\mathcal{S}_s = - \left(\frac{\partial F}{\partial T} \right)_{V,N} \tag{4.245}$$

and also the superconducting heat capacity at constant volume

$$C_s = T\left(\frac{\partial S}{\partial T}\right)_{V,N} = -T\left(\frac{\partial^2 F}{\partial T^2}\right)_{V,N} \tag{4.246}$$

with $V = L^D$ the D-dimensional volume of the system, and N is the total number of superconducting electrons. Actually N does not appear explicitly in the formula of the superconducting free energy.

4.9 Josephson Junctions

Two superconductors or superfluids separated by a tunneling barrier give rise to the so-called Josephson junction, originally predicted by Brian Josephson in 1962. In contrast to superconducting Josephson junctions, it is possible to have a huge population imbalance with atomic Josephson junctions due to the appearance of the self-trapping phenomena. In this section we study a Josephson junction by using the real-time Feynman path-integral approach. In particular, we consider a system of interacting bosons that are tunneling between two sites. At the mean-field (saddle-point) level, we obtain a phase-imbalance model. Then, performing path integration over the population imbalance, we derive an only-phase effective action.

4.9.1 Two-Site Model

The simplest real-time Lagrangian of a system made of bosonic particles which are tunneling between two sites ($j = 1, 2$) is given by

$$L = \sum_{j=1,2}\left[i\hbar\,\psi_j^*\dot{\psi}_j + \mu_j|\psi_j|^2 - \frac{U}{2}|\psi_j|^4\right] + \frac{J}{2}\left(\psi_1^*\psi_2 + \psi_2^*\psi_1\right) \tag{4.247}$$

where $\psi_j(t)$ is the adimensional complex field of bosons in the j site at real time t, μ_j is the chemical potential on the j site, U is the on-site interaction strength of particles (supposed to be the same of each site), J is the tunneling energy, $\hbar$ is the reduced Planck constant, i is the imaginary unit, and dot means the derivative with respect to time t. In the case of a superconducting junction the chemical potential difference $\Delta\mu = \mu_2 - \mu_1$ takes into account an applied bias voltage.

This Lagrangian can describe the macroscopic quantum tunneling of a system of identical bosonic alkali-metal atoms of mass m. The Lagrangian is also used to model the tunneling of Cooper pairs of electrons with opposite spin under the simpliyfing assumption that these Cooper pairs can be treated as bosonic-like particles with mass $m = 2m_e$, with m_e is the mass of one electron.

To clarify the role of the hopping term, which contains the tunneling energy J, we perform the Madelung transformation

$$\psi_j(t) = \sqrt{N_j(t)}\, e^{i\phi_j(t)} , \tag{4.248}$$

where $N_j(t)$ is the number of bosons in the j site and $\phi_j(t)$ is the phase angle. We also introduce the total number

$$N = N_1(t) + N_2(t) \tag{4.249}$$

which is a constant of motion, and the relative phase

$$\phi(t) = \phi_2(t) - \phi_1(t), \tag{4.250}$$

which is not a constant of motion, similarly to the total phase

$$\phi_{tot}(t) = \phi_1(t) + \phi_2(t) . \tag{4.251}$$

We also define the population imbalance as

$$z(t) = \frac{N_1(t) - N_2(t)}{N} . \tag{4.252}$$

In this way the Lagrangian can be rewritten as

$$L(\phi, z) = \frac{N\hbar}{2} z\dot{\phi} - \frac{UN^2}{4} z^2 + \frac{JN}{2}\sqrt{1-z^2}\cos(\phi) - \frac{N\Delta\mu}{2} z , \tag{4.253}$$

that is the phase-imbalance Lagrangian of a Josephson junction. Notice that we have safely removed constant terms and also the exact differential $\dot{\phi}_{tot}$.

4.9.2 Mean-Field Dynamics

A complete quantum mechanical description of our Josephson junction model can be derived from the real-time Feynman path integral

$$\int \mathcal{D}[\psi_1^*, \psi_1]\mathcal{D}[\psi_2^*, \psi_2] \; e^{\frac{i}{\hbar}\int_{t_I}^{t_F} dt\, L(\psi_1,\psi_1^*,\psi_2,\psi_2^*)} = \int \mathcal{D}[\phi]\mathcal{D}[z] \; e^{\frac{i}{\hbar}\int_{t_I}^{t_F} dt\, L(\phi,z)} \, , \quad (4.254)$$

integrating over all the configurations of the dynamical variables $\phi(t)$ and $z(t)$. The calculation of these path integrals is a very difficult task, also numerically. To get analytical results, some approximation is needed.

The simplest approximation scheme to treat our quantum problem is the mean-field (saddle-point) approximation, where one takes into account only the configurations that extremize the action functional

$$S = \int_{t_I}^{t_F} dt \, L(\phi, z) \, . \quad (4.255)$$

These configurations are the ones that satisfy the Euler-Lagrange equations. In our case, they are

$$\hbar\dot{\phi} = J\frac{z}{\sqrt{1-z^2}}\cos(\phi) + UNz + \Delta\mu \, , \quad (4.256)$$

$$\hbar\dot{z} = -J\sqrt{1-z^2}\sin(\phi) \, . \quad (4.257)$$

These equations describe the mean-field dynamics of the macroscopic quantum tunneling in a Josephson junction, where $\phi(t)$ is the relative phase angle of the complex field of the Bose-Einstein condensate (or superfluid or superconductor) between the two junctions at time t and $z(t)$ is the corresponding relative population imbalance of the Bose condensed particles (or Cooper pairs).

Adopting the BCS theory, Josephson obtained a simplified version of Eqs. (4.256) and (4.257), namely

$$\hbar\dot{\phi} = \Delta\mu \, , \quad (4.258)$$

$$\hbar\dot{z} = -J\sin(\phi) \, , \quad (4.259)$$

because he assumed a small population imbalance, i.e., $|z(t)| \ll 1$, a negligible on-site interaction, i.e., $U \approx 0$, and also a quite small tunneling energy, i.e., $J|z| \ll \hbar \dot{z}$. Introducing the (neutral) current $I(t) = -N\dot{z}(t)/\hbar$ and the critical current $I_c = NJ/\hbar$, we have

$$\dot{\phi} = \frac{\Delta\mu}{\hbar} , \tag{4.260}$$

$$I = I_c \sin(\phi) . \tag{4.261}$$

These are the so-called Josephson equations of superconductors. In the absence of a bias $\Delta\mu$, one gets a constant relative phase, i.e., $\phi(t) = \phi_0$, and the corrent reads

$$I = I_c \sin(\phi_0) . \tag{4.262}$$

This equation is the direct current (DC) Josephson effect: in the absence of an applied bias voltage, there is nevertheless a current due to the existence of a constant phase difference ϕ_0. Instead, in the presence of a bias $\Delta\mu$, we have

$$I = I_c \sin\left(\frac{\Delta\mu}{\hbar} t\right) , \tag{4.263}$$

that is the alternating current (AC) Josephson effect. This is an amazing phenomenon: contrary to Ohmic circuits, a constant voltage induces in Josephson circuits an alternate current. Josephson was awarded the Nobel Prize in Physics in 1973 for his prediction of these DC and AC effects in superconductors, which were quickly confirmed by experiments. Let us consider again to Eq. (4.253), working in the absence of the chemical potential bias, i.e., $\Delta\mu = 0$ but without the very strong assumptions that $U \approx 0$ and $J|z| \ll \hbar\dot{z}$. We assume however that both $\phi(t)$ and $z(t)$ are small, i.e., $|\phi(t)| \ll 1$ and $|z(t)| \ll 1$. In this way the Lagrangian (4.253) can be approximated as

$$L^{(2)} = \frac{N\hbar}{2} z\dot{\phi} - \frac{JN}{4}\phi^2 - \frac{(JN + UN^2)}{4} z^2 , \tag{4.264}$$

removing a constant term. The Euler-Lagrange equations of this quadratic Lagrangians are

$$\hbar\dot{\phi} = (J + UN)z , \tag{4.265}$$

$$\hbar\dot{z} = -J\phi , \tag{4.266}$$

which can be rewritten as a single equation for the harmonic oscillation of $\phi(t)$ and the harmonic oscillation of $z(t)$, given by

$$\ddot{\phi} + \Omega^2 \, \phi = 0 \, , \tag{4.267}$$

$$\ddot{z} + \Omega^2 \, z = 0 \, , \tag{4.268}$$

both with frequency

$$\Omega = \frac{1}{\hbar}\sqrt{J^2 + NUJ} \, , \tag{4.269}$$

that is the mean-field frequency of macroscopic quantum oscillation in terms of tunneling energy J, interaction strength U, and number N of particles. The validity of Eq. (4.269) has been confirmed experimentally with atomic Bose-Einstein condensates in a quasi one-dimensional configuration and confined in a double-well potential.

4.9.3 Effective Only-Phase Action

The quantum-mechanical way to derive an effective only-phase action S_0 for $\phi(t)$, starting from the full action $S[\phi(t), z(t)]$, is to trace out the dynamical variable $z(t)$ with a path integral over it. However, the Lagrangian L of Eq. (4.253) cannot be used to extract analytically, this effective only-phase Lagrangian because one can explicitly calculate only quadratic integrals. To perform these calculations we can use quadratic expansions, i.e., the Gaussian approximation. For the sake of simplicity, we set $\Delta\mu = 0$. Expanding the Lagrangian L of Eq. (4.253) at the Gaussian level with respect to $z(t)$, we obtain

$$L(\phi, z) = \frac{N\hbar}{2} z\dot{\phi} - \frac{UN^2 + JN\cos(\phi)}{4} z^2 + \frac{JN}{2}\cos(\phi) \, . \tag{4.270}$$

Still the problem is not easy to solve because of the coupling between $\cos(\phi)$ and z^2. Here, we consider a further simplification, given by $UN \gg J$, which implies

$$L(\phi, z) = \frac{N\hbar}{2} z\dot{\phi} - \frac{UN^2}{4} z^2 + \frac{JN}{2}\cos(\phi) \, . \tag{4.271}$$

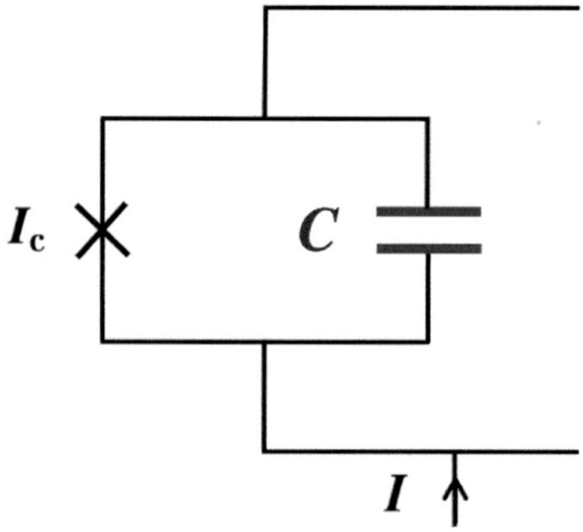

Figure 4.3
Josephson junction, denoted by $\times$, coupled with an electrical capacitance C. Here, I_c denotes the critical current in the Josephson junction, and I is the external current.

Now the Gaussian integration with respect to z can be done extending the codomain of $z(t)$ from $[-1, 1]$ to $]-\infty, +\infty[$. One then finds

$$\int \mathcal{D}[\phi]\mathcal{D}[z]\; e^{\frac{i}{\hbar}\int dt\, L(\phi,z)} = \int \mathcal{D}[\phi]\; e^{iS_J[\phi]/\hbar}\,, \tag{4.272}$$

where

$$S_J[\phi] = \int_{t_I}^{t_F} dt \left[\frac{\hbar^2\dot{\phi}^2}{4U} + \frac{JN}{2}\cos(\phi)\right] \tag{4.273}$$

is the effective only-phase action. This effective action describes quite well the capacitively-shunted superconducting Josephson junction, where the population imbalance is very small, i.e., $|z(t)| \ll 1$, but the number of boson-like Cooper pairs N is large.

Within the framework of superconducting junctions, $eI_c = eJN/\hbar$ is the critical electric current with e the electric charge of the proton, and $C = e^2/(2U)$ is the electric capacitance. In fact, for the superconducting Josephson junction, we have $\Delta\mu = eV$ with V, the electric potential energy difference. From Eq. (4.260) it follows

$$V = \frac{\hbar}{e}\dot{\phi}\,. \tag{4.274}$$

For a capacitor we have instead

$$V = \frac{Q}{C} \, . \tag{4.275}$$

If the two devices are in parallel, their potential energy difference must be the same. Then

$$\frac{\hbar}{e}\dot{\phi} = \frac{Q}{C} \tag{4.276}$$

or

$$\frac{\hbar}{e}\ddot{\phi} = -\frac{eI}{C} = -\frac{eI_c}{C}\sin(\phi) \tag{4.277}$$

using $\dot{Q} = -eI$. Thus

$$\ddot{\phi} + \frac{e^2 I_c}{\hbar C}\sin(\phi) = 0 \, . \tag{4.278}$$

This is indeed the Euler-Lagrange equation of the only-phase Lagrangian

$$L(\phi) = \frac{\hbar^2 C}{2e^2}\dot{\phi}^2 + \frac{\hbar I_c}{2}\cos(\phi) \, . \tag{4.279}$$

In conclusion we stress that superconducting Josephson junctions are currently used as building blocks in several quantum computers. In these devices, two quantum states of the Josephson junctions are used as superconducting qubits. Nowadays, many companies, universities and research institutes are working in the field superconducting quantum computing, developing quantum chips which uses a superconducting architecture.

4.10 Solved Problems

Problem 4.1

For a non-interacting and uniform two-spin-component Fermi gas in 3D dimension, the chemical potential at zero temperature (Fermi energy) is given by

$$\mu = \frac{\hbar^2}{2m}(3\pi^2)^{2/3} n^{2/3}$$

where n is the number density. From this formula calculate the internal energy E.

Solution

At zero temperature the internal energy density $\mathcal{E} = E/L^3$ is related to the chemical potential μ by

$$\mathcal{E} = \int_0^n \mu(\tilde{n})\, d\tilde{n} \ .$$

In our problem

$$\begin{aligned}\mathcal{E} &= \int_0^n \frac{\hbar^2}{2m}(3\pi^2)^{2/3}\tilde{n}^{2/3}\, d\tilde{n} = \frac{\hbar^2}{2m}(3\pi^2)^{2/3}\int_0^n \tilde{n}^{2/3}\, d\tilde{n} \\ &= \frac{\hbar^2}{2m}(3\pi^2)^{2/3}\frac{3}{5}n^{5/3} \ .\end{aligned}$$

Problem 4.2

Let us consider the Grassmann variables c and $\bar{c}$. Taking into account the anti-commutation properties of the Grassmann numbers and the defining properties of the Brezin integral, calculate

$$\int d\bar{c}\, dc\ e^{-\bar{c}Ac}$$

and also

$$\int d\bar{c}\, dc\ e^{\bar{c}Ac}$$

with A is a complex number.

Solution

Due to the fact that we are considering Grasmann numbers, it follows that

$$e^{-\bar{c}Ac} = 1 - \bar{c}Ac \ ,$$

because all the other terms of the Taylor-MacLaurin expansion are identically zero. Consequently

$$\int d\bar{c}\, dc\ e^{-\bar{c}Ac} = \int d\bar{c}\, dc\ (1 - \bar{c}Ac) \ .$$

Finally, on the basis of the definition of the Brezin integral, where the integration is defined as equivalent to (left) differentiation, we get

$$\int d\bar{c}\, dc\ (1 - \bar{c}Ac) = \frac{\partial^2}{\partial\bar{c}\partial c}(1 - \bar{c}Ac) = \frac{\partial^2}{\partial\bar{c}\partial c}(-\bar{c}Ac) = A\frac{\partial^2}{\partial\bar{c}\partial c}(c\bar{c})\,.$$

Thus

$$\int d\bar{c}\, dc\ e^{-\bar{c}Ac} = A\ .$$

Following the same reasoning, we have

$$\int d\bar{c}\, dc\ e^{\bar{c}Ac} = \int d\bar{c}\, dc\ \ (1 + \bar{c}Ac)\ \ .$$

Then

$$\int d\bar{c}\, dc\ (1 + \bar{c}Ac) = \frac{\partial^2}{\partial\bar{c}\partial c}(1 + \bar{c}Ac) = \frac{\partial^2}{\partial\bar{c}\partial c}(\bar{c}Ac) = -A\frac{\partial^2}{\partial\bar{c}\partial c}(c\bar{c})$$

and finally

$$\int d\bar{c}\, dc\ e^{\bar{c}Ac} = -A\ .$$

Problem 4.3

Prove that

$$\prod_{\mathrm{odd}\, n=-\infty}^{+\infty} (-i\,\pi\, n + x)\, e^{i\pi n 0^+} = (1 + e^{-x})\ ,$$

where $e^{i\pi n 0^+}$ is the convergence factor.

Solution

First of all, we set

$$\mathcal{R}(x) = \prod_{\mathrm{odd}\, n=-\infty}^{+\infty} (-i\,\pi\, n + x)\, e^{i\pi n 0^+}\ .$$

Then we write

$$\begin{aligned}\ln(\mathcal{R}(x)) &= \sum_{\text{odd}\, n=-\infty}^{+\infty} [\ln(-i\,\pi\,n + x) + i\pi n 0^+] \\ &= \sum_{\text{odd}\, n=-\infty}^{+\infty} \ln(-i\,\pi\,n + x)\, e^{i\pi n 0^+} \,.\end{aligned}$$

Its derivative with respect to x gives

$$\frac{d}{dx}\ln(\mathcal{R}(x)) = \sum_{\text{odd}\, n=-\infty}^{+\infty} \frac{e^{i\pi n 0^+}}{-i\,\pi\,n + x} \,.$$

We also set

$$\mathcal{T}(x) = \sum_{\text{odd}\, n=-\infty}^{+\infty} \frac{e^{i\pi n 0^+}}{-i\,\pi\,n + x} \,.$$

Following a procedure similar to the one discussed in Problem 3.4, we now consider the function

$$J_\delta(x) = \frac{1}{2\pi i}\oint_{\mathcal{C}} dz\, \frac{e^{z\delta}}{-z + x}\, \frac{1}{e^z + 1}$$

with $\mathcal{C}$ a very large (infinite) circle around the origin $z = 0$ in the Argand-Gauss plane of the complex variable z, we realize that the integrating function has poles at $z = x$ but also at $z = i\pi n$ with odd n. The corresponding residues are $e^{x\delta}/(e^x+1)$ and $1/(-i2\pi n + x)$. Moreover, this function is such that

$$\lim_{\delta\to 0^+} J_\delta(x) = 0$$

because, due to the presence of the term $e^{z\delta}$, the integrand function is exponentially suppressed in the full Argand-Gauss plane as $z \to \infty$. From the residue theorem, we also have

$$\lim_{\delta\to 0^+} J_\delta(x) = \mathcal{T}(x) - \frac{1}{e^x + 1} \,.$$

Thus, comparing the last two equations we get

$$\mathcal{T}(x) = \frac{1}{e^x + 1} \,.$$

Remembering that

$$\mathcal{T}(x) = \frac{d}{dx} \ln(\mathcal{R}(x))$$

it follows

$$\ln(\mathcal{R}(x)) = \int \mathcal{T}(x) dx + K = \ln\big(1 + e^{-x}\big) + K$$

with K a constant of integration. Removing the (infinite) constant K, we finally obtain

$$\mathcal{R}(x) = (1 + e^{-x}) \, .$$

Problem 4.4
Prove that

$$\prod_{\text{odd}\, n=-\infty}^{+\infty} (-i\,\pi\, n + x) = 2 \cosh\!\left(\frac{x}{2}\right) .$$

Solution
We set

$$\mathcal{R}(x) = \prod_{\text{odd}\, n=-\infty}^{+\infty} (-i\,\pi\, n + x) \, .$$

Then we find

$$\ln(\mathcal{R}(x)) = \sum_{\text{odd}\, n=-\infty}^{+\infty} \ln(-i\,\pi\, n + x)$$

and

$$\frac{d}{dx} \ln(\mathcal{R}(x)) = \sum_{\text{odd}\, n=-\infty}^{+\infty} \frac{1}{-i\,\pi\, n + x}$$

We also set

$$\mathcal{T}(x) = \sum_{\text{odd}\, n=-\infty}^{+\infty} \frac{1}{-i\,\pi\, n + x} \, .$$

The fraction can be separated into real and imaginary parts as follows

$$\frac{1}{-i\pi n + x} = \frac{x}{\pi^2 n^2 + x^2} + i\frac{\pi n}{\pi^2 n^2 + x^2}$$

but after the summation over n, the imaginary part is zero. As a consequently we can write

$$\mathcal{T}(x) = x \sum_{\text{odd } n=-\infty}^{+\infty} \frac{1}{\pi^2 n^2 + x^2} = x \sum_{n=-\infty}^{+\infty} \frac{1}{\pi^2(2n+1)^2 + x^2} .$$

The series converges, and it is given by

$$\sum_{n=-\infty}^{+\infty} \frac{1}{\pi^2(2n+1)^2 + x^2} = \frac{1}{2x} \tanh\left(\frac{x}{2}\right) .$$

Thus, we find

$$\mathcal{T}(x) = \frac{1}{2} \tanh\left(\frac{x}{2}\right) .$$

Recalling that

$$\mathcal{T}(x) = \frac{d}{dx} \ln(\mathcal{R}(x))$$

it follows that

$$\ln(\mathcal{R}(x)) = \int \mathcal{T}(x) dx + K = \ln\left(2\cosh\left(\frac{x}{2}\right)\right) + K$$

with K a constant of integration. Removing K by hand, we finally obtain

$$\mathcal{R}(x) = 2 \cosh\left(\frac{x}{2}\right) .$$

Problem 4.5

Starting from the only-phase action

$$S_J[\phi] = \int_{t_I}^{t_F} dt \left[\frac{\hbar^2 \dot{\phi}^2}{4U} + \frac{JN}{2} \cos(\phi) \right]$$

of the bosonic Josephson junction, derive the quantum propagator $K(\phi_F, t_F | \phi_I, t_I)$ under the assumption of small phase fluctuations around $\phi = 0$.

Solution

We perform the second-order Taylor expansion

$$\cos(\phi) \simeq 1 - \frac{1}{2}\phi^2$$

and the approximated quadratic action reads

$$S_J[\phi] \simeq \int_{t_I}^{t_F} \left[\frac{\hbar^2 \dot{\phi}^2}{4U} - \frac{JN}{4}\phi^2 \right] dt + \frac{JN}{2}(t_F - t_I) \ .$$

This is the action of a harmonic oscillator with effective mass

$$M = \frac{\hbar^2}{2U}$$

and effective frequency

$$\Omega = \sqrt{\frac{NUJ}{\hbar^2}} \ .$$

As a consequence, the quantum propagator is that one of this harmonic oscillator:

$$\begin{aligned} K(\phi_F, t_F|\phi_I, t_I) \ &\simeq \ \sqrt{\frac{M\Omega}{2\pi i\hbar \sin(\Omega(t_F - t_I))}} \, e^{\frac{i}{\hbar}\frac{JN}{2}(t_F - t_I)} \\ &\times \ e^{\frac{i}{\hbar}\frac{M\Omega}{2}\left[(\phi_F^2 + \phi_I^2)\cot(\Omega(t_F - t_I)) - \frac{2\phi_F\phi_I}{\sin(\Omega(t_F - t_I))}\right]} \ . \end{aligned}$$

Further Reading

For the functional integration of the non-relativistic bosonic and fermionic fields:
– N. Nagaosa, Quantum Field Theory in Condensed Matter Physics, chapt. 1, sections 1.1 and 1.2 (Springer, Berlin, 1999).
– H.T.C. Stoof, K.B. Gubbels, and D.B.M. Dickerscheid, Ultracold Quantum Fields, chapt. 6, sections 6.1, 6.2, 6.3, and 6.4 (Springer, Berlin, 2009).
– A. Altland and B. Simons, Condensed Matter Field Theory, chapt. 2, section 2.1 (Cambridge Univ. Press, Cambridge, 2006).
For the dimensional regularization of Gaussian fluctuations in the case of interacting bosons and fermions:
– L. Salasnich and F. Toigo, Zero-point energy of ultracold atoms, Phys. Rep. **640**, 1 (2016).
– T. Papenbrock and G. F. Bertsch, Phys. Rev. C **59**, 2052 (1999).
For the BCS theory in metals with various experimental data:
– N.W. Ashcroft and N.D. Mermin, Solid State Physics (Holt-Sounders, Philadelphia, 1976).
– J.P. Carbotte, Properties of boson-exchange superconductors, Rev. Mod. Phys. **62**, 1027 (1990).
For the condensate fraction of superfluid fermions:
– L. Salasnich, N. Manini, and A. Parola, Condensate Fraction of a Fermi Gas in the BCS-BEC Crossover, Phys. Rev. A **72**, 023621 (2005).
For a detailed account of the Ginzburg-Landau theory and its developements:
– A. Larkin and A. Varlamov, Theory of Fluctuations in Superconductors (Oxford Univ. Press, Oxford, 2009).
In discussing Josephson junnctions we have closely followed the paper
– K. Furutani, J. Tempere, and L. Salasnich, Quantum effective action for the bosonic Josephson junction, Phys. Rev. B **105**, 134510 (2022).
Two-dimensional fermionic superfluid and superconductors are discussed in these three papers:

– L Salasnich, Condensate fraction of a two-dimensional attractive Fermi gas, Phys. Rev. A **76**, 015601 (2007).
– G Bighin and L Salasnich, Finite-temperature quantum fluctuations in two-dimensional Fermi superfluids, Physical Review B **93**, 014519 (2016).
– Y. Nakagawa, Y. Kasahara, T. Nomoto, R. Arita, T. Nojima, and Y. Iwasa, Gate-controlled BCS-BEC crossover in a two-dimensional superconductor, Science **372**, 190 (2021).
– M. Lanaro, G. Bighin, L. Dell'Anna, and L. Salasnich, Finite-size effects in the two-dimensional BCS-BEC crossover, Phys. Rev. B **109**, 104511 (2024).

5

Path Integral for Photons

In this chapter we consider the quantum statistical mechanics of identical relativistic massless bosons, i.e., the photons, by introducing the formalism of functional integration for the electromagnetic field. In this way we derive the familiar Planck's law of the black body radiation. We also consider three enlightening applications: the Casimir effect, the thermodynamics of photons in nonlinear Kerr media, and the Bose-Einstein condensation of photons in a optical microcavity.

5.1 Introduction

The beginning of quantum mechanics is set in 1900 when Max Planck found that the only way to explain the experimental results of the electromagnetic spectrum emitted by hot solid bodies is to assume that the energy E of the radiation with angular frequency ω emitted from the walls of the body is quantized according to

$$E = \hbar\omega \, n \, , \tag{5.1}$$

where $n = 0, 1, 2, ...$ is an integer quantum number and $\hbar$ is the reduced Planck constant. With the help of statistical mechanics Planck derived the following expression

$$\rho(\omega) = \frac{\hbar}{\pi^2 c^3} \frac{\omega^3}{e^{\hbar\omega/(k_B T)} - 1} \tag{5.2}$$

for the electromagnetic energy density per unit of frequency $\rho(\omega)$ emitted by the hot body at the temperature T, with k_B the Boltzmann constant and c is the speed of light in a vacuum. This formula, known as Planck's law of the black-body radiation, is in

DOI: 10.1201/9781003588337-5

very good agreement with experimental data. Note that a hot solid body can indeed be approximated by the so-called black body, that is an idealized physical body that absorbs all incident electromagnetic radiation, and it is also the best possible emitter of thermal radiation.

We now show that the Planck law can be deduced by applying the path integral formalism to the electromagnetic field seen as a set of infinite classical harmonic oscillators, each oscillator corresponding to a different frequency of the electromagnetic spectrum.

5.2 Partition Function of Photons

For a system of relativistic non-interacting massless photons, the chemical potential is identically zero, i.e., $\mu = 0$. In fact, the chemical potential μ of a many-body system is the energy required to add a particle to the system. The minimal energy to create a particle of mass m from the vacuum is nothing else than its rest energy mc^2. Thus, for a gas of non-interacting photons, we have $\mu = 0$ and consequently the number of photons is not fixed. Thus, the grand canonical partition function of photons is given by

$$\mathcal{Z} = \mathrm{Tr}[e^{-\beta \hat{H}}] , \tag{5.3}$$

where the second-quantized Hamiltonian is

$$\hat{H} = \int_V d^3\mathbf{r} \left\{ \frac{\epsilon_0}{2}(\partial_t \hat{\mathbf{A}})^2 + \frac{1}{2\mu_0}(\boldsymbol{\nabla} \wedge \hat{\mathbf{A}})^2 \right\} = \sum_{\mathbf{k}s} \hbar\omega_k \; \hat{a}^+_{\mathbf{k}s}\hat{a}_{\mathbf{k}s} \tag{5.4}$$

working with the Coulomb gauge in a container of volume V. Remember that $\omega_k = ck$ and $\beta = 1/(k_B T)$; moreover $\hat{\mathbf{A}}(\mathbf{r}, t)$ is the electromagnetic vector potential operator, such that

$$\hat{\mathbf{A}}(\mathbf{r}, t) = \sum_{\mathbf{k}s} \sqrt{\frac{\hbar}{2\varepsilon_0 \omega_k}} \left[\hat{a}_{\mathbf{k}s}(t) \, \frac{e^{i\mathbf{k}\cdot\mathbf{r}}}{\sqrt{V}} + \hat{a}^+_{\mathbf{k}s}(t) \, \frac{e^{-i\mathbf{k}\cdot\mathbf{r}}}{\sqrt{V}} \right] \mathbf{u}_{\mathbf{k}s} \; , \tag{5.5}$$

where $\hat{a}_{\mathbf{k}s}(t)$ and $\hat{a}^+_{\mathbf{k}s}(t)$ are the annihilation and creation operators of photons with wavevector $\mathbf{k}$ and polarization $s = 1, 2$.

Following the procedure discussed in the initial part of Chapter 3, the partition function (5.3) can be reformulated as a path integral by using bosonic coherent states. Doing so we have

$$\mathcal{Z} = \int \mathcal{D}[\mathbf{A}(\mathbf{r},\tau)] \; \exp\left\{ -\frac{S_E[\mathbf{A}(\mathbf{r},\tau)]}{\hbar} \right\} , \tag{5.6}$$

where

$$\int \mathcal{D}[A(\mathbf{r},\tau)] = \int \prod_{(\mathbf{r},\tau)} d\mathbf{A}(\mathbf{r},\tau) \tag{5.7}$$

is the path-integral measure over all the possible classical vectors potentials and

$$S_E[\mathbf{A}(\mathbf{r},\tau)] = \int_0^{\beta\hbar} d\tau \int_V d^3\mathbf{r} \left\{ \frac{\epsilon_0}{2}(\partial_t \mathbf{A})^2 - \frac{1}{2\mu_0}(\boldsymbol{\nabla} \wedge \mathbf{A})^2 \right\} \tag{5.8}$$

is the Euclidean action functional in terms of the vector potential field. Alternatively, we can consider the representation of the quantum Hamiltonian (5.4) in terms of ladder operators. In this way we find

$$\mathcal{Z} = \int \mathcal{D}[\{a_{\mathbf{k}s}\},\{a^*_{\mathbf{k}s}\}] \; \exp\left\{ -\frac{S_E[\{a_{\mathbf{k}s}\},\{a^*_{\mathbf{k}s}\}]}{\hbar} \right\} , \tag{5.9}$$

where

$$\int \mathcal{D}[\{a^*_{\mathbf{k}s}\},\{a_{\mathbf{k}s}\}] = \int \prod_{\tau} \prod_{\mathbf{k}s} \frac{da^*_{\mathbf{k}s}(\tau) da_{\mathbf{k}s}(\tau)}{2\pi i} \tag{5.10}$$

is the measure over the complex numbers, $a_{\mathbf{k}s}(\tau)$ and $a^*_{\mathbf{k}s}(\tau)$ which are the "classical" dynamical variables of the problem in imaginary time τ, with Euclidean action

$$S_E[\{a^*_{\mathbf{k}s}\},\{a_{\mathbf{k}s}\}] = \int_0^{\beta\hbar} d\tau \left\{ \sum_{\mathbf{k}s} a^*_{\mathbf{k}s}(\tau) \left(\hbar\partial_\tau + \hbar\omega_k\right) a_{\mathbf{k}s}(\tau) \right\} . \tag{5.11}$$

Introducing the Fourier representation of the dynamical variables

$$a_{\mathbf{k}s}(\tau) = \sum_n a_{\mathbf{k}s,n} \, e^{-i\omega_n \tau} \tag{5.12}$$

with

$$\omega_n = \frac{2\pi n}{\beta\hbar} \tag{5.13}$$

the bosonic Matsubara frequencies, the Euclidean action can be written in diagonal form as

$$S_E[\{a^*_{\mathbf{k}s,n}\}, \{a_{\mathbf{k}s,n}\}] = \hbar \sum_{\mathbf{k}sn} \lambda_{k,n} |a_{\mathbf{k}s,n}|^2 \, . \tag{5.14}$$

where

$$\lambda_{k,n} = \beta \left(-i\hbar\omega_n + \hbar\omega_k \right) \, . \tag{5.15}$$

Using the new variables, the partition function can be explicitly calculated because it involves Gaussian integrals. We have

$$\begin{aligned} \mathcal{Z} &= \int \mathcal{D}[\{a^*_{\mathbf{k}s,n}\}, \{a_{\mathbf{k}s,n}\}] \, \exp\left\{ -\sum_{\mathbf{k}sn} \lambda_{k,n} |a_{\mathbf{k}s,n}|^2 \right\} \\ &= \frac{1}{\prod_{\mathbf{k}sn} \lambda_{k,n}} = \frac{1}{\prod_{\mathbf{k}s} (1 - e^{-\beta\hbar\omega_k})} \, . \end{aligned} \tag{5.16}$$

See Chapter 3 for the analog case of massive non-relativistic bosons.

5.2.1 Gas of Photons at Thermal Equilibrium

The grand potential $\Omega = F - TS - \mu N$ coincides with the Helmholtz free energy $F = E - TS$ because $\mu = 0$, and it is given by

$$\Omega = -\frac{1}{\beta} \ln(\mathcal{Z}) = \frac{1}{\beta} \sum_{\mathbf{k}s,n} \ln\left(\lambda_{k,n}\right) = \frac{1}{\beta} \sum_{\mathbf{k}s} \ln\left(1 - e^{-\beta\hbar\omega_k}\right) \, . \tag{5.17}$$

The internal energy can be derived as

$$E = \langle \hat{H} \rangle = \frac{1}{\mathcal{Z}} Tr[\hat{H} \, e^{-\beta\hat{H}}] = -\frac{\partial}{\partial\beta} \ln\left(\mathcal{Z}\right) \tag{5.18}$$

After some calculations, one finds

$$E = \sum_{\mathbf{k}s} \frac{\hbar\omega_k}{e^{\beta\hbar\omega_k} - 1} = \sum_{\mathbf{k}s} \hbar\omega_k \, N_{\mathbf{k}s} \tag{5.19}$$

where

$$N_{\mathbf{k}s} = \langle \hat{N}_{\mathbf{k}s} \rangle = \frac{1}{e^{\beta\hbar\omega_k} - 1} \tag{5.20}$$

is the thermal average of the number of photons with wavevector $\mathbf{k}$ and polarization s. In the continuum limit, where

$$\sum_{\mathbf{k}} \to V \int \frac{d^3\mathbf{k}}{(2\pi)^3}, \tag{5.21}$$

with V the volume, and taking into account that $\omega_k = ck$, removing the zero-point energy, E_{vac} of vacuum, one can write the energy density $\mathcal{E} = \langle \hat{H} \rangle / V$ as

$$\mathcal{E} = 2 \int \frac{d^3\mathbf{k}}{(2\pi)^3} \frac{c\hbar k}{e^{\beta c\hbar k} - 1} = \frac{c\hbar}{\pi^2} \int_0^\infty dk \frac{k^3}{e^{\beta c\hbar k} - 1}, \tag{5.22}$$

where the factor 2 is due to the two possible polarizations ($s = 1, 2$). By using $\omega = ck$ instead of k as the integration variable, one gets

$$\mathcal{E} = \frac{\hbar}{\pi^2 c^3} \int_0^\infty d\omega \frac{\omega^3}{e^{\beta\hbar\omega} - 1} = \int_0^\infty d\omega \, \rho(\omega), \tag{5.23}$$

where

$$\rho(\omega) = \frac{\hbar}{\pi^2 c^3} \frac{\omega^3}{e^{\beta\hbar\omega} - 1} \tag{5.24}$$

is the energy density per frequency, i.e., the familiar formula of the black-body radiation, obtained for the first time by Max Planck. The previous integral can be explicitly calculated, and it gives

$$\mathcal{E} = \frac{\pi^2 k_B^4}{15 c^3 \hbar^3} T^4, \tag{5.25}$$

which is nothing but the Stefan-Boltzmann law. Similarly one determines the average number density of photons:

$$\bar{n} = \frac{\langle \hat{N} \rangle}{V} = \frac{1}{\pi^2 c^3} \int_0^\infty d\omega \, \frac{\omega^2}{e^{\beta\hbar\omega} - 1} = \frac{2\zeta(3) k_B^3}{\pi^2 c^3 \hbar^3} T^3, \tag{5.26}$$

where $\zeta(3) \simeq 1.202$. Both the energy density $\mathcal{E}$ and the number density n of photons go to zero as the temperature T goes to zero.

5.3 Casimir Effect

In the treatment of the previous section, we have not included the zero-point energy E_{vac} of the electromagnetic field, that is given by

$$E_{vac} = \sum_{\mathbf{k}} \sum_{s} \frac{1}{2} \hbar c \sqrt{k_x^2 + k_y^2 + k_z^2} = V \int \frac{d^3\mathbf{k}}{(2\pi)^3} \hbar c \sqrt{k_x^2 + k_y^2 + k_z^2} \tag{5.27}$$

by using the dispersion relation $\omega_k = ck = c\sqrt{k_y^2 + k_y^2 + k_z^2}$. In particular, considering a region with the shape of a parallelepiped of length L, along both x and y and length ℓ along z, the volume V is given by $V = L^2\ell$, and the vacuum energy E_{vac} in the region is

$$\begin{aligned} E_{vac} &= \hbar c \int_{-\infty}^{+\infty} \frac{L\, dk_x}{2\pi} \int_{-\infty}^{+\infty} \frac{L\, dk_y}{2\pi} \int_{-\infty}^{+\infty} \frac{\ell\, dk_z}{2\pi} \sqrt{k_x^2 + k_y^2 + k_z^2} \\ &= \frac{\hbar c}{2\pi} L^2 \int_0^{\infty} dk_{\parallel} k_{\parallel} \left[\int_0^{\infty} dn \sqrt{k_{\parallel}^2 + \frac{n^2\pi^2}{\ell^2}} \right] , \end{aligned} \tag{5.28}$$

where the second expression is obtained setting $k_{\parallel} = \sqrt{k_x^2 + k_y^2}$ and $n = (\ell/\pi)k_z$.

Let us now consider the presence of two perfect metallic plates with the shape of a square of length L having parallel faces lying in the (x, y) plane at distance ℓ. Along the z-axis the stationary standing waves of the electromagnetic field vanishes on the metal plates and the k_z component of the wavevector $\mathbf{k}$ is no more a continuum variables but it is quantized via

$$k_z = n\frac{\pi}{\ell} , \tag{5.29}$$

where now $n = 0, 1, 2, ...$ is an integer number, and not a real number as in Eq. (5.28). In this case the zero-point energy in the volume $V = L^2\ell$ between the two plates reads

$$E'_{vac} = \frac{\hbar c}{2\pi} L^2 \int_0^{\infty} dk_{\parallel} k_{\parallel} \left[\frac{k_{\parallel}}{2} + \sum_{n=1}^{\infty} \sqrt{k_{\parallel}^2 + \frac{n^2\pi^2}{\ell^2}} \right] , \tag{5.30}$$

where for $k_z = 0$ the state $|(k_z, k_y, 0)1\rangle$ with polarization $\boldsymbol{\epsilon}_{\mathbf{k}1}$ parallel to the z axis (and orthogonal to the metallic plates) is bound, while the state $|(k_z, k_y, 0)2\rangle$ with polarization $\boldsymbol{\epsilon}_{\mathbf{k}2}$ orthogonal to the z axis (and parallel to the metallic plates) is not bound, and it does not contribute to the discrete summation.

The difference between E'_{vac} and E_{vac} divided by L^2 gives the net energy per unit surface area $\mathcal{E}$, namely

$$\mathcal{E} = \frac{E'_{vac} - E_{vac}}{L^2} = \frac{\hbar c}{2\pi} \left(\frac{\pi}{\ell}\right)^3 \left[\frac{1}{2}A(0) + \sum_{n=1}^{\infty} A(n) - \int_0^{\infty} dn\, A(n)\right] , \tag{5.31}$$

where we have defined

$$A(n) = \int_0^{+\infty} d\zeta\, \zeta\, \sqrt{\zeta^2 + n^2} = \frac{1}{3}\left[(n^2 + \infty)^{3/2} - n^2\right] , \tag{5.32}$$

with $\zeta = (\ell/\pi)k_{\|}$. Notice that $A(n)$ is clearly divergent but Eq. (5.31) is not divergent due to the cancellation of divergences with opposite sign. By using the Euler-MacLaurin formula for the difference of infinite series and integrals, one eventually obtains

$$\mathcal{E} = -\frac{\pi^2}{720} \frac{\hbar c}{\ell^3} . \tag{5.33}$$

From this energy difference, $\mathcal{E}$, one deduces that there is an attractive force per unit area $\mathcal{F}$, between the two plates, given by

$$\mathcal{F} = -\frac{d\mathcal{E}}{d\ell} = -\frac{\pi^2}{240} \frac{\hbar c}{\ell^4} . \tag{5.34}$$

Numerically this result, predicted in 1948 by Hendrik Casimir during his research activity at the Philips Physics Laboratory in Eindhoven, is very small

$$\mathcal{F} = -\frac{1.30 \cdot 10^{-27}\text{N m}^2}{\ell^4} . \tag{5.35}$$

It has been experimentally verified for the first time by Steven Lamoreaux in 1997. There are now several experimental confirmations of this amazing effect.

5.4 Thermal Photons in Kerr Media

In Chapter 1 we have seen that the QFT Hamiltonian of a gas of photons in nonlinear Kerr media can be written as

$$\hat{H} = \sum_{\mathbf{k}s} \left[\hbar v_1^{(0)} k \ \hat{N}_{\mathbf{k}s} + \hbar v_1^{(1)} k \ \hat{N}_{\mathbf{k}s} \ \hat{N}_{\mathbf{k}s} \right] , \tag{5.36}$$

where $v_1^{(0)}$ is the speed of light in these media and $v_1^{(1)}$ is a coefficient taking into account the photon-photon interaction. The Hamiltonian can be simplified by using the Hartree-Fock approximation, i.e.

$$\hat{N}_{\mathbf{k}s} \ \hat{N}_{\mathbf{k}s} \simeq 2\langle \hat{N}_{\mathbf{k}s} \rangle \ \hat{N}_{\mathbf{k}s} - \langle \hat{N}_{\mathbf{k}s} \rangle^2 , \tag{5.37}$$

obtaining

$$\begin{aligned} \hat{H} &= \sum_{\mathbf{k}s} \left[\left(\hbar v_1^{(0)} k + 2\hbar v_1^{(1)} k \ \langle \hat{N}_{\mathbf{k}s} \rangle \right) \hat{N}_{\mathbf{k}s} - \hbar v_1^{(1)} k \ \langle \hat{N}_{\mathbf{k}s} \rangle^2 \right] \qquad (5.38) \\ &= \sum_{\mathbf{k}s} E_k \hat{a}^+_{\mathbf{k}s} \hat{a}_{\mathbf{k}s} - \sum_{\mathbf{k}s} \hbar v_1^{(1)} k \ \langle \hat{N}_{\mathbf{k}s} \rangle^2 . \end{aligned}$$

with

$$E_k = \hbar v_1^{(0)} k + 2\hbar v_1^{(1)} k \ \langle \hat{N}_{\mathbf{k}s} \rangle . \tag{5.39}$$

Notice that $\langle \hat{N}_{\mathbf{k}s} \rangle$ does not depend on s. The partition function is then given by

$$\mathcal{Z} = e^{\beta \sum_{\mathbf{k}s} \hbar v_1^{(1)} k \ \langle \hat{N}_{\mathbf{k}s} \rangle^2} \int \mathcal{D}[\{a_{\mathbf{k}s,n}\}, \{a^*_{\mathbf{k}s,n}\}] \ \exp \left\{ - \sum_{\mathbf{k}sn} \lambda_{k,n} |a_{\mathbf{k}s,n}|^2 \right\} \tag{5.40}$$

with

$$\lambda_{k,n} = \beta \left(-i\hbar\omega_n + E_k \right) . \tag{5.41}$$

After performing the Gaussian integrals, we obtain the grand potential

$$\Omega = -\frac{1}{\beta} \ln(\mathcal{Z}) = \frac{1}{\beta} \sum_{\mathbf{k}s} \left[-\hbar v_1^{(1)} k \ \langle \hat{N}_{\mathbf{k}s} \rangle^2 + \ln\left(1 - e^{-\beta E_k}\right) \right] . \tag{5.42}$$

5.5 Bose-Einstein Condensation of Photons

We have seen several times that the dispersion relation of photons is given by

$$\omega = ck = c\sqrt{k_x^2 + k_y^2 + k_z^2} \ . \tag{5.43}$$

For photons confined in an optical microcavity the component k_z of the wavevector $\mathbf{k} = (k_x, k_y, k_z)$ can be fixed. As a consequence, for small k_x and k_y, we have

$$\hbar\omega = E_z + \frac{\hbar^2}{2m^*}(k_x^2 + k_y^2) + \ldots \ , \tag{5.44}$$

where

$$E_z = \hbar c|k_z| \tag{5.45}$$

and

$$m^* = \hbar|k_z|^{1/2} \tag{5.46}$$

introducing a sort of effective mass m^* for the photons. Quite remarkably, having fixed $k_z \neq 0$, the energy spectrum is gapped and the minimal energetic cost to create a photon is E_{00}. The photons in the optical microcavity have a transverse Gaussian-like shape, which is well captured by modeling the photons under the effect of a transverse harmonic confinement

$$U(x, y) = \frac{1}{2}m^*\omega_\perp^2(x^2 + y^2) \ , \tag{5.47}$$

with $\omega_\perp$ the frequency which giving the characteristic length of the transverse harmonic confinement

$$\ell_\perp = \sqrt{\frac{\hbar}{m^*\omega_\perp}} \ . \tag{5.48}$$

Thus the total energy of a single photon is given by

$$E = E_z + \frac{p_x^2 + p_y^2}{2m^*} + \frac{1}{2}m^*\omega_\perp^2(x^2 + y^2) \tag{5.49}$$

where $p_k = \hbar k_x$ and $p_y = \hbar k_y$. In 2010 Klaers *et al.* measured for the first time a Bose–Einstein condensate (BEC) of photons in this system. The cavity mirrors provide both a confining potential and a non-vanishing effective photon mass, making the system formally equivalent to a two-dimensional gas of trapped massive bosons. In the experiment the photons thermalize to the room temperature of the dye solution by multiple scattering with the dye molecules. The single-particle quantized energy levels associated with the classical energy of Eq. (5.49) are

$$E_{n_x n_y} = E_{00} + \hbar\omega_\perp (n_x + n_y) \tag{5.50}$$

with quantum numbers $n_x, n_y = 0, 1, 2, 3, ...$ and

$$E_{00} = E_z + \hbar\omega_\perp = \hbar c|k_z| + \hbar\omega_\perp \ . \tag{5.51}$$

Considering the bosonic Matsubara frequencies (5.13), and taking into account the crucial role of a non-vanishing chemical potential μ, the Euclidean action of our problem reads

$$\begin{aligned} S_E[\{a^*_{n_x n_y s,n}\}, \{a_{n_x n_y s,n}\}] = \hbar \sum_{n_x n_y s n} \beta\left(-i\hbar\omega_n + E_{n_x n_y} - \mu\right) \\ \times \left|a_{n_x n_y s,n}\right|^2 , \end{aligned} \tag{5.52}$$

where $a_{n_x n_y s,n}$ are the complex coefficients associated with the coherent states of photons with transverse quantum numbers n_x and n_y, polarization s, and Matsubara index n. The grand canonical partition function is then

$$\begin{aligned} \mathcal{Z} &= \int \mathcal{D}[\{a^*_{n_x n_y s,n}\}, \{a_{n_x n_y s,n}\}] \\ &\quad \exp\left\{-S_E[\{a^*_{n_x n_y s,n}\}, \{a_{n_x n_y s,n}\}]/\hbar\right\} \\ &= \frac{1}{\prod_{n_x n_y s n} \beta\left(-i\hbar\omega_n + E_{n_x n_y} - \mu\right)} \ . \end{aligned} \tag{5.53}$$

It follows this grand potential

$$\Omega = -\frac{1}{\beta}\ln(\mathcal{Z}) = \frac{1}{\beta}\sum_{n_x n_y s} \ln\left(1 - e^{-\beta(E_{n_x n_y} - \mu)}\right) \tag{5.54}$$

and this average number of photons

$$N = \langle \hat{N} \rangle = -\frac{\partial \Omega}{\partial \mu} = 2 \sum_{n_x n_y} \frac{1}{e^{\beta(E_{n_x n_y} - \mu)} - 1} \; . \tag{5.55}$$

In the BEC phase, where $\mu = E_{00}$, we get

$$N = N_0 + 2 \sum_{(n_x n_y) \neq (0,0)} \frac{1}{e^{\beta \hbar \omega_\perp (x_x + n_y)} - 1} \tag{5.56}$$

where $N_0 = 2N_{00}$ is the number of photons in the condensate taking into account the two polarizations. In the continuum limit

$$\sum_{(n_x n_y) \neq (0,0)} \to \int_{-\infty}^{+\infty} dn_x \int_{-\infty}^{+\infty} dn_y \tag{5.57}$$

the critical temperature T_{BEC} of Bose-Einstein condensation is obtained setting N_{00}, from which

$$N = 2 \int_{-\infty}^{+\infty} dn_x \int_{-\infty}^{+\infty} dn_y \frac{1}{e^{\hbar \omega_\perp (x_x + n_y)/(k_B T_{BEC})} - 1} \; . \tag{5.58}$$

The two integrals can be evaluated analytically, and we have

$$k_B T_{BEC} = \frac{\sqrt{3N}}{\pi} \hbar \omega_\perp \; . \tag{5.59}$$

Finally, we also derive the condensate fraction, that is

$$\frac{N_0}{N} = 1 - \left(\frac{T}{T_{BEC}} \right)^2 \; . \tag{5.60}$$

It is relevant to emphasize that the 2D gas of photons is not uniform and consequently, the Mermin-Wagner theorem does not hold.

5.6 Solved Problems

Problem 5.1

An optical cavity is made of two mirrors with the electromagnetic field confined in between. Assume a quasi one-dimensional monochromatic light with a fixed mirror at the position $x = 0$. The other mirror, with mass m and position is $x = q(t)$, can instead move. Determine the equation of motion of the coordinate $q(t)$ of the moving mirror.

Solution

The real-time Lagrangian of the system under investigation can be written as

$$L = L_q + L_A \ ,$$

where

$$L_q = \frac{m}{2}\dot{q}^2$$

is the Lagrangian of the moving mirror with mass m, while

$$L_A = \frac{\epsilon_0\, S}{2} \int_0^q dx \left[\left(\frac{\partial A}{\partial t} \right)^2 - c^2 \left(\frac{\partial A}{\partial x} \right)^2 \right]$$

is the Lagrangian of the electromagnetic field with vector potential $A(x,t)$ with S the transverse area of the two mirrors. The two Lagrangians are coupled due to the upper limit of integration in L_A. A simple variational ansatz for the field $A(x,t)$ is

$$A(x,t) = Q(t) \sqrt{\frac{2}{q(t)}} \sin\left(\frac{\pi\, l}{q(t)} x \right).$$

This is the very special case of a monochromatic electromagnetic wave, where only one single mode, with the quantum number l, wavenumber $k_l(t) = \pi l / q(t)$ and amplitude $Q(t)$, is occupied. This single-mode ansatz satisfies the boundary conditions

$$A(0,t) = A(q(t),t) = 0$$

and $Q(t)$ is the time-dependent variational parameter. Inserting the variational ansatz into L_A and integrating over x, we obtain the effective Lagrangian

$$L = \frac{m}{2}\dot{q}^2 + \frac{\epsilon_0 S}{2} \left(\dot{Q}^2 - \frac{c^2 \pi^2 l^2}{q^2} Q^2 \right) .$$

neglecting the coupling terms between $\dot{q}(t)$ and $Q(t)$. The corresponding Euler-Lagrange equations are

$$\begin{aligned} \frac{m\,\ddot{q}}{\left(1-\frac{\dot{q}^2}{c^2}\right)^{3/2}} - \epsilon_0 S \frac{c^2\pi^2 l^2}{q^3} Q^2 &= 0\,, \\ \ddot{Q} + \frac{c^2\pi^2 l^2}{q^2} Q &= 0\,. \end{aligned}$$

Problem 5.2

Determine the following thermal average of the number correlator

$$\langle \hat{N}_{\mathbf{k}s} \hat{N}_{\mathbf{k}'s'} \rangle_T = \frac{1}{\mathcal{Z}} Tr[\hat{N}_{\mathbf{k}s} \hat{N}_{\mathbf{k}'s'}\, e^{-\beta \hat{H}}]\,.$$

Solution

To obtain this thermal average, we introduce a fictitious partition function

$$\mathcal{Z}_F = Tr[e^{-\beta \hat{H}_F}]$$

associated to the fictitious Hamiltonian

$$\hat{H}_F = \sum_{\mathbf{k}s} E_{\mathbf{k}s}\, \hat{N}_{\mathbf{k}s}$$

where $E_{\mathbf{k}s}$ are fictitious single-mode energies. It is straightforward to show that

$$\mathcal{Z}_F = \prod_{\mathbf{k}s} \frac{1}{1 - e^{-\beta E_{\mathbf{k}s}}}$$

and clearly

$$\ln(\mathcal{Z}_F) = -\sum_{\mathbf{k}s} \ln\left(1 - e^{-\beta E_{\mathbf{k}s}}\right)\,.$$

Moreover, we have

$$\begin{aligned} \langle \hat{N}_{\mathbf{k}s} \rangle_T &= \frac{1}{\mathcal{Z}_F} Tr[\hat{N}_{\mathbf{k}s}\, e^{-\beta \hat{H}_F}] = \frac{1}{\mathcal{Z}_F} Tr[\hat{N}_{\mathbf{k}s}\, e^{-\beta \sum_{\mathbf{k}'s'} E_{\mathbf{k}'s'} \hat{N}_{\mathbf{k}'s'}}] \\ &= -\frac{1}{\beta} \frac{1}{\mathcal{Z}_F} \frac{\partial}{\partial E_{\mathbf{k}s}} \mathcal{Z}_F = -\frac{1}{\beta} \frac{\partial}{\partial E_{\mathbf{k}s}} \ln(\mathcal{Z}_F)\,. \end{aligned}$$

One can calculate the partial derivative with respect to $E_{\mathbf{k}',s'}$. After some lengthy manipulations, one gets

$$\langle \hat{N}_{\mathbf{k}s} \hat{N}_{\mathbf{k}'s'} \rangle_T = \langle \hat{N}_{\mathbf{k}s} \rangle_T \, \langle \hat{N}_{\mathbf{k}'s'} \rangle_T + \delta_{\mathbf{k},\mathbf{k}'} \, \delta_{s,s'} \, \langle \hat{N}_{\mathbf{k}s} \rangle_T \left(1 + \langle \hat{N}_{\mathbf{k}s} \rangle_T \right) ,$$

where $\delta_{i,j}$ is the Kronecker delta. Finally, setting $E_{\mathbf{k}s} = \hbar\omega_k$ and $\beta = 1/(k_B T)$, we obtain explicitly

$$\begin{aligned} \langle \hat{N}_{\mathbf{k}s} \hat{N}_{\mathbf{k}'s'} \rangle_T \; &= \; \frac{1}{\left(e^{\frac{\hbar\omega_k}{k_B T}} - 1 \right) \left(e^{\frac{\hbar\omega_{k'}}{k_B T}} - 1 \right)} \\ &\quad + \delta_{\mathbf{k},\mathbf{k}'} \, \delta_{s,s'} \, \frac{1}{e^{\frac{\hbar\omega_k}{k_B T}} - 1} \left(1 + \frac{1}{e^{\frac{\hbar\omega_k}{k_B T}} - 1} \right) . \end{aligned}$$

Problem 5.3

Calculate the thermal average number $\langle \hat{N}_{\mathbf{k}s} \rangle$ of photons in a Kerr medium under the assumption of a very low temperature.

Solution

In the very low temperature regime, we have

$$\langle \hat{N}_{\mathbf{k}s} \rangle = \frac{1}{e^{\beta(\hbar v_1^{(0)} k + 2\hbar v_1^{(1)} k \langle \hat{N}_{\mathbf{k}s} \rangle)} - 1} \simeq \frac{k_B T}{\hbar v_1^{(0)} k + 2\hbar v_1^{(1)} k \langle \hat{N}_{\mathbf{k}s} \rangle}$$

because $(e^x - 1) \simeq (1 + x + ... - 1) = x + ...$ It follows

$$2(hv_1^{(1)} k) \, \langle \hat{N}_{\mathbf{k}s} \rangle^2 + (\hbar v_1^{(0)} k) \, \langle \hat{N}_{\mathbf{k}s} \rangle - k_B T \simeq 0$$

and finally

$$\langle \hat{N}_{\mathbf{k}s} \rangle \simeq \frac{-(\hbar v_1^{(0)} k) + \sqrt{(\hbar v_1^{(0)} k)^2 + 8(\hbar v_1^{(1)} k) k_B T}}{4(\hbar v_1^{(1)} k)} .$$

Problem 5.4

Find the internal energy $E = \langle \hat{H} \rangle$ of the BEC of photons in a microcavity.

Solution

In full generality, within the Grand Canonical ensemble, we have

$$\begin{aligned}\langle \hat{H} - \mu \hat{N} \rangle &= \frac{1}{\mathcal{Z}} Tr[(\hat{H} - \mu \hat{N}) e^{-\beta(\hat{H} - \mu \hat{N})}] \\ &= -\frac{1}{\mathcal{Z}} \left(\frac{\partial}{\partial \beta} Tr[e^{-\beta(\hat{H} - \mu \hat{N})}] \right)_\mu \\ &= -\frac{1}{\mathcal{Z}} \left(\frac{\partial \mathcal{Z}}{\partial \beta} \right)_\mu = - \left(\frac{\partial}{\partial \beta} (\ln \mathcal{Z}) \right)_\mu ,\end{aligned}$$

and consequently, the internal energy E can be written as

$$E = \langle \hat{H} \rangle = \langle \hat{H} - \mu \hat{N} \rangle + \mu \langle \hat{N} \rangle = - \left(\frac{\partial}{\partial \beta} (\ln \mathcal{Z}) \right)_\mu + \mu \langle \hat{N} \rangle .$$

In our specific problem of bosons in 2D harmonic oscillator with frequency $\omega_\perp$, we have

$$- \ln \mathcal{Z} = \ln (1 - e^{\beta(\mu - \hbar\omega_\perp)}) + \frac{1}{(\beta\hbar\omega_\perp)^2} J(e^{\beta(\mu - \hbar\omega_\perp)}) ,$$

where

$$J(x) = \int_0^{+\infty} dy\, y \ln\big(1 - x e^{-y}\big) .$$

Thus, we have a formula for the internal energy E as a function of the chemical potential μ and the temperature $T = 1/(k_B \beta)$, i.e., $E = E(\mu, T)$. Or, equivalently, the internal energy E as a function of β and the fugacity $z = e^{\beta\mu}$, i.e., $E = E(\beta, z)$. The constraint

$$N = \langle \hat{N} \rangle = \frac{e^{\beta(\mu - \hbar\omega_\perp)}}{1 - e^{\beta(\mu - \hbar\omega_\perp)}} + \frac{1}{(\beta\hbar\omega_\perp)^2} I(e^{\beta(\mu - \hbar\omega_\perp)}) ,$$

where

$$I(x) = \int_0^{+\infty} dy \frac{y}{x^{-1} e^y - 1} ,$$

gives the total number of particles as a function of the chemical potential μ and the temperature $T = 1/(k_B\beta)$, i.e., $N = N(\mu, T)$. Or, equivalently, the total number N as a function of β and the fugacity z, i.e., $N = N(\beta, z)$. In the BEC phase, where $0 \leq T \leq T_{BEC}$ with

$$k_B T_{BEC} = \frac{1}{\sqrt{I(1)}} \hbar\omega_\perp N^{1/2}$$

and $\mu \to \hbar\omega_\perp$, we obtain

$$\frac{E}{\hbar\omega_\perp} = N + 2\zeta(3) \left(\frac{k_B T}{\hbar\omega_\perp} \right)^3 ,$$

with $J(1) = -\zeta(3) = -1.20206$ and $\zeta(x)$ the Riemann zeta function. Rewriting the formula in terms of T, N, and T_c one easily finds

$$\frac{E}{N k_B T_{BEC}} = \sqrt{I(1)} \left[\frac{1}{N^{1/2}} + \frac{2\zeta(3)}{I(1)^{3/2}} \left(\frac{T}{T_{BEC}} \right)^3 \right] .$$

Finally, in the regime $N \gg 1$ the previous relation becomes

$$\frac{E}{N k_B T_{BEC}} = \frac{2\zeta(3)}{I(1)} \left(\frac{T}{T_{BEC}} \right)^3 .$$

Problem 5.5

Find the Schwarzschild radius r_s of a black hole of mass M from the escape velocity of a photon. Then, derive the Hawking temperature T of the black hole considering the creation-annihilation of virtual photons at the event horizon of radius r_s.

Solution

In celestial mechanics the escape velocity, or the escape speed, is the minimum speed needed for an object to escape from the orbit of a primary body. Classical mechanics tells us that the escape velocity v_{esc} of a particle from a gravitational body of mass M is given by

$$v_{\text{esc}} = \sqrt{\frac{2GM}{r}} ,$$

where r is the distance between the particle and the celestial body. A photon in a vacuum always has the speed c. From

$$v_{\text{esc}} = c$$

we get

$$\sqrt{\frac{2GM}{r}} = c \, .$$

Identifying r with the Schwarzschild radius r_s, we then obtain

$$r_s = \frac{2GM}{c^2} \, .$$

This formula was obtained for the first time in 1916 by Karl Schwarzschild calculatted the exact solution of the Einstein field equations of general relativity for the gravitational field of a non-rotating, spherically symmetric body with mass M.

We now consider the creation of a pair of virtual photons moving with opposite direction at the distance r_s from the center of the black hole, assuming that one photon escapes while the other remains trapped in the black hole. The minimum of energy-time uncertainty principle tells us that

$$\Delta E \, \Delta t = \frac{\hbar}{2}$$

with $\hbar$ the familiar reduced Planck constant. We suppose that the energy ΔE is setting the thermal energy scale, namely

$$\Delta E = k_B T$$

with k_B the familiar Boltzmann constant. A natural timescale Δt is instead the time it takes the light to travel around the horizon, given by

$$\Delta t = \frac{2\pi r_s}{c} \, .$$

In this way we get

$$k_B T \frac{2\pi r_s}{c} = \frac{\hbar}{2} \, .$$

Using the formula obtained for r_s, we finally find

$$T = \frac{\hbar c^3}{8\pi G M k_B} \, ,$$

that is the Hawking temperature, derived for the first time in 1974 by Stephen Hawking with a more rigorous reasoning.

Further Reading

For the functional integration of the electromagnetic field:
– N. Nagaosa, Quantum Field Theory in Condensed Matter Physics, chapt. 1, sections 1.1 and 1.2 (Springer, Berlin, 1999).
– A. Altland and B. Simons, Condensed Matter Field Theory, chapt. 2, section 2.1 (Cambridge Univ. Press, Cambridge, 2006).
For the gas of photons at finite temperature:
– K. Huang, Statistical Mechanics, chap. 12, section 12.1 (Wiley, New York, 1987).
For the Casimir effect:
– R.W. Robinett, Quantum Mechanics: Classical Results, Modern Systems, and Visualized Examples, chap. 19, section 19.9 (Oxford Univ. Press, Oxford, 2006).
– S. K. Lamoreaux, Demonstration of the Casimir force in the 0.6 to 6 μm range, Phys. Rev. Lett. **78**, 5 (1997).
– G. Bressi, G. Carugno, R. Onofrio, and G. Ruoso, Measurement of the Casimir force between parallel metallic surfaces, Phys. Rev. Lett. **88**, 041804 (2002).
For the Bose-Einstein condensate of photons in a microcavity:
– J. Klaers, J. Schmitt, F. Vewinger, and M. Weitz, Bose–Einstein condensation of photons in an optical microcavity, Nature **468**, 545 (2010).

Bibliography

[1] A. Altland and B. Simons. *Condensed Matter Field Theory.* Cambridge Univ. Press, Cambridge, 2006.

[2] F.T. Arecchi. Measurement of the statistical distribution of gaussian and laser sources. *Phys. Rev. Lett.*, 15:912, 1965.

[3] V.I. Arnold. *Mathematical Methods of Classical Mechanics.* Springer, Berlin, 1989.

[4] N.W. Ashcroft and N.D. Mermin. *Solid State Physics.* Holt-Sounders, Philadelphia, 1976.

[5] M. Le Bellac. *A Short Introduction to Quantum Information and Quantum Computation.* Cambridge Univ. Press, Cambridge, 2006.

[6] V.B. Berestetskii, E.M. Lifshitz, and L.P. Pitaevskii. *Relativistic Quantum Theory, vol. 4 of Course of Theoretical Physics.* Pergamon Press, Oxford, 1971.

[7] G. Bighin and L. Salasnich. Finite-temperature quantum fluctuations in two-dimensional fermi superfluids. *Physical Review B*, 93:014519, 2016.

[8] J.D. Bjorken and S.D. Drell. *Relativistic Quantum Mechanics.* McGraw-Hill, New York, 1964.

[9] B.H. Bransden and C.J. Joachain. *Physics of Atoms and Molecules.* Prentice Hall, Upper Saddle River, 2003.

[10] J.P. Carbotte. Properties of boson-exchange superconductors. *Rev. Mod. Phys.*, 62:1027, 1990.

[11] N. Dupuis. *Field Theory of Condensed Matter and Ultracold Gases, vol. 1.* World Scientific, Singapore, 2023.

[12] A. Scott (ed.). *Encyclopedia of Nonlinear Science.* Taylor and Francis, New York, 2005.

[13] J.V. Jose (ed.). *40 Years Of Berezinskii Kosterlitz Thouless Theory.* World Scientific, Singapore, 2013.

[14] W. Zwerger (ed.). *The BCS-BEC Crossover and the Unitary Fermi Gas.* Lecture Notes in Physics. Springer, Berlin, 2012.

[15] A.L. Fetter and J.D. Walecka. *Quantum Theory of Many-Particle Systems.* Dover Publications, New York, 2003.

[16] R.P. Feynman, A.R. Hibbs, and D.F. Styer. *Quantum Mechanics and Path Integrals. Emended Edition.* Dover, New York, 2010.

[17] K. Furutani, J. Tempere, and L. Salasnich. Quantum effective action for the bosonic josephson junction. *Phys. Rev. B*, 105:134510, 2022.

[18] H. Goldstein, C.P. Poole, and J. Safko. *Classical Mechanics.* Pearson, London, 2011.

[19] K. Huang. *Statistical Mechanics.* Wiley, New York, 1987.

[20] J. Klaers, J. Schmitt, F. Vewinger, and M. Weitz. Bose–einstein condensation of photons in an optical microcavity. *Nature*, 468:545, 2010.

[21] H. Kleinert. *Path Integrals In Quantum Mechanics, Statistics, Polymer Physics, And Financial Markets.* World Scientific, Singapore, 2009.

[22] G. Kordas, S.I. Mistakidis, and A.I. Karanikas. Coherent-state path integrals in the continuum. *Phys. Rev. A*, 90:032104, 2014.

[23] S.K. Lamoreaux. Demonstration of the casimir force in the 0.6 to 6 m range. *Phys. Rev. Lett.*, 78:5, 1997.

[24] M. Lanaro, G. Bighin, L. Dell'Anna, and L. Salasnich. Finite-size effects in the two-dimensional bcs-bec crossover. *Phys. Rev. B*, 109:104511, 2024.

[25] A. Larkin and A. Varlamov. *Theory of Fluctuations in Superconductors.* Oxford Univ. Press, Oxford, 2009.

[26] U. Leonhardt. *Measuring the Quantum State of Light.* Cambridge Univ. Press, Cambridge, 1997.

[27] F. Mandl and G. Shaw. *Quantum Field Theory.* Wiley, New York, 1984.

[28] G. Mazzarella, L. Salasnich, A. Parola, and F. Toigo. Coherence and entanglement in the ground-state of a bosonic josephson junction:from macroscopic schrödinger cats to separable fock states. *Phys. Rev. A*, 83:053607, 2011.

[29] N. Nagaosa. *Quantum Field Theory in Condensed Matter Physics.* Springer, Berlin, 1999.

[30] Y. Nakagawa, Y. Kasahara, T. Nomoto, R. Arita, T. Nojima, and Y. Iwasa. Gate-controlled bcs-bec crossover in a two-dimensional superconductor. *Science*, 372:190, 2021.

[31] J.W. Negele and H. Orland. *Quantum Many-particle Systems.* Addison-Wesley, Boston, 1988.

[32] M. Ogilvie. *A Multidisciplinary Approach to Quantum Field Theory. An introduction.* IOP Publishing, Bristol, 2022.

[33] A. Rancon. Hubbard-stratonovich transformation and consistent ordering in the coherent state path integral: insights from stochastic calculus. *J. Phys. A: Math. Theor.*, 53:105302, 2020.

[34] W. Rindler. *Introduction to Special Relativity.* Oxford Univ. Press, Oxford, 1991.

[35] L. Salasnich. Condensate fraction of a two-dimensional attractive fermi gas. *Phys. Rev. A*, 76:015601, 2007.

[36] L. Salasnich. *Quantum Physics of Light and Matter: Photons, Atoms, and Strongly Correlated Systems.* Springer, Cham, 2017.

[37] L. Salasnich, N. Manini, and A. Parola. Condensate fraction of a fermi gas in the bcs-bec crossover. *Phys. Rev. A*, 72:023621, 2005.

[38] L. Salasnich, A. Parola, and L. Reatto. Effective wave equations for the dynamics of cigar-shaped and disk-shaped bose condensates. *Phys. Rev. A*, 65:043614, 2002.

[39] L. Salasnich and F. Toigo. Zero-point energy of ultracold atoms. *Phys. Rep.*, 640:1, 2016.

[40] M.O. Scully and M.S. Zubairy. *Quantum Optics.* Cambridge Univ. Press, Cambridge, 1997.

[41] H.T.C. Stoof, K.B. Gubbels, and D.B.M. Dickerscheid. *Ultracold Quantum Fields.* Springer, Berlin, 2009.

[42] J. Zinn-Justin. *Quantum Field Theory and Critical Phenomena.* Oxford Univ. Press, 2021.

Index

For Product Safety Concerns and Information please contact our EU representative GPSR@taylorandfrancis.com Taylor & Francis Verlag GmbH, Kaufingerstraße 24, 80331 München, Germany